PROTEIN BIOCHEMISTRY, SYNTHESIS, STRUCTURE AND CELLULAR FUNCTIONS

PROTEIN – WATER INTERACTIONS

A DIFFERENTIAL APPROACH

AF675925

PROTEIN BIOCHEMISTRY, SYNTHESIS, STRUCTURE AND CELLULAR FUNCTIONS

Additional books in this series can be found on Nova's website under the Series tab.

Additional e-books in this series can be found on Nova's website under the e-book tab.

PROTEIN BIOCHEMISTRY, SYNTHESIS, STRUCTURE AND CELLULAR FUNCTIONS

PROTEIN – WATER INTERACTIONS

A DIFFERENTIAL APPROACH

VLADIMIR A. SIROTKIN
EDITOR

NOVINKA

New York

Copyright © 2014 by Nova Science Publishers, Inc.

All rights reserved. No part of this book may be reproduced, stored in a retrieval system or transmitted in any form or by any means: electronic, electrostatic, magnetic, tape, mechanical photocopying, recording or otherwise without the written permission of the Publisher.

For permission to use material from this book please contact us:
Telephone 631-231-7269; Fax 631-231-8175
Web Site: http://www.novapublishers.com

NOTICE TO THE READER

The Publisher has taken reasonable care in the preparation of this book, but makes no expressed or implied warranty of any kind and assumes no responsibility for any errors or omissions. No liability is assumed for incidental or consequential damages in connection with or arising out of information contained in this book. The Publisher shall not be liable for any special, consequential, or exemplary damages resulting, in whole or in part, from the readers' use of, or reliance upon, this material. Any parts of this book based on government reports are so indicated and copyright is claimed for those parts to the extent applicable to compilations of such works.

Independent verification should be sought for any data, advice or recommendations contained in this book. In addition, no responsibility is assumed by the publisher for any injury and/or damage to persons or property arising from any methods, products, instructions, ideas or otherwise contained in this publication.

This publication is designed to provide accurate and authoritative information with regard to the subject matter covered herein. It is sold with the clear understanding that the Publisher is not engaged in rendering legal or any other professional services. If legal or any other expert assistance is required, the services of a competent person should be sought. FROM A DECLARATION OF PARTICIPANTS JOINTLY ADOPTED BY A COMMITTEE OF THE AMERICAN BAR ASSOCIATION AND A COMMITTEE OF PUBLISHERS.

Additional color graphics may be available in the e-book version of this book.

Library of Congress Cataloging-in-Publication Data

ISBN: 978-1-63463-007-8

Library of Congress Control Number: 2014950567

Published by Nova Science Publishers, Inc. † New York

CONTENTS

PREFACE

This book is aimed at understanding which molecular parameters control the thermodynamics, structure, and functions of the protein-water systems. Proteins are one of the most important classes of biological molecules. Water binding (hydration or biological water) plays a crucial role in determining the structure, stability, and functions of proteins. Knowledge of processes occurring upon hydration or dehydration of protein macromolecules is very important in biotechnological and pharmaceutical applications of proteins such as their use as biocatalysts, biosensors, and selective adsorbents.

There are essential differences between hydration and bulk water surrounding a protein. This means that a characterization of the hydration of protein macromolecules requires elucidating the effects of both the protein on water and *vice versa*. Therefore, a quantitative estimation of the protein and water contributions to the thermodynamic functions of binary protein-water systems is of considerable fundamental importance and practical interest.

This book describes the basic principles of a novel methodology to investigate the protein-water interactions. This methodology is based on the analysis of the excess thermodynamic functions of mixing. The thermodynamic properties (volume V, enthalpy H, entropy S, heat capacity C_p, and Gibbs free energy G) of a real binary water-protein system can be expressed in terms of the excess functions. They are the difference between the thermodynamic function of mixing in a real system and the value corresponding to an ideal system at the same temperature, pressure and composition. For an ideal system, all excess functions are zero. Deviations of the excess functions from zero indicate the extent to which the studied binary system is non-ideal due to strong specific interactions between components (i.e., hydrogen bonding and charge-charge interactions).

The advantages of our methodology are:

(i) We are able to simultaneously determine the excess partial quantities of water and proteins;
(ii) These thermodynamic quantities can be determined in the entire range of water content.

In the present book, high-precision densitometry, differential scanning calorimetry, and isothermal calorimetry were applied to characterize the hydration dependencies of the excess thermodynamic functions of binary protein-water systems.

Ovalbumin, ribonuclease A, lysozyme, chymotrypsinogen A, serum albumin, lactoglobulin were used as model proteins. These biomacromolecules represent a series of proteins in which the hydrophobicity of proteins is gradually changed in a wide range.

Chapters 1-5 describe the excess thermodynamic functions of the binary systems obtained as a function of composition at 25 ^{o}C. The aim of these studies is to demonstrate how these quantities correlate with the coverage of the protein macromolecules by water molecules.

Chapters 6 and 7 present the analysis of the hydration and structure of proteins in the presence of organic solvents by isothermal calorimetry and FTIR spectroscopy.

In: Protein – Water Interactions
Editor: Vladimir A. Sirotkin

ISBN: 978-1-63463-007-8
© 2014 Nova Science Publishers, Inc.

Chapter 1

Analysis of Hydration of Binary Protein-Water Mixtures. Methodology

Vladimir A. Sirotkin*
Kazan Federal University, A. M. Butlerov
Institute of Chemistry, Kazan, Russia

Abstract

This chapter describes the basic principles of a novel methodology to investigate the protein-water interactions. This methodology is based on the analysis of the excess thermodynamic functions (volume V, enthalpy H, entropy S, heat capacity C_p, and Gibbs free energy G) of mixing. The advantages of our methodology are: i) We are able to simultaneously determine the excess partial quantities of water and proteins; ii) These thermodynamic quantities can be determined in the entire range of water content.

Keywords: Protein (biomacromolecule) hydration, Methodology, Binary mixtures, Excess functions

* Corresponding author: Vladimir A. Sirotkin. Kazan Federal University, A. M. Butlerov Institute of Chemistry, Kremlevskaya street, 18, Kazan, 420008, Russia. E-mail: vsir@mail.ru.

1. INTRODUCTION

Water binding (hydration or biological water) plays a key role in determining the structure, stability, dynamics, and functions of proteins [1-7]. On the other hand, there are essential differences between hydration and the bulk water surrounding a protein. Thus, a characterisation of the hydration of proteins requires elucidating the effects of the protein on water and *vice versa*.

Therefore, a quantitative estimation of the protein and water contributions to the thermodynamic functions of binary protein-water systems is of considerable importance and practical interest.

All the thermodynamic quantities are some derivatives of Gibbs free energy, *G*, with respect to a variable or a combination thereof. The Gibbs energy, *G*, the zeroth order derivative, dictates the fate of the protein-water systems. For example, the first differentiation with respect to temperature, *T*, separates the entropy and the enthalpy effects. Further differentiation with respect to *T* gives the heat capacity, *C*p. The higher the order of the derivative, the more detailed the information that it contains.

The measured thermodynamic properties of the binary water-protein systems (volume *V*, enthalpy *H*, entropy *S*, heat capacity C_p, and Gibbs free energy *G*) [1-3, 8-13] contain total information, including the corresponding conformational changes in the protein structure and the glass transition. However, the simultaneous estimation of the protein and water contributions in the entire range of water content has not been attempted.

In this chapter, we present a novel methodology to investigate the protein-water interactions [14-17]. This methodology is based on the analysis of the excess thermodynamic functions of mixing. It facilitates the individual evaluation of the protein and water partial quantities in the entire range of water content. The present study is aimed at understanding which molecular parameters control the excess thermodynamic functions of binary protein-water systems.

2. METHODOLOGY

2.1. Excess Partial Quantities

The thermodynamic properties (volume *V*, enthalpy *H*, entropy *S*, and Gibbs free energy *G*) of a real binary water-protein system can be expressed in terms of the excess functions, Z^E. They are the difference between the thermodynamic function of mixing in a real system and the value corresponding to an ideal system at the same temperature, pressure and composition. For an ideal system, all excess functions are zero. Deviations of the excess functions from zero indicate the extent to which the studied binary system is non-ideal due to strong specific interactions between components (i.e., hydrogen bonding and charge-charge interactions). More comprehensive reviews of excess functions are given in References [18-20].

The Z^E values can be calculated using Equation 1:

$$Z^E = Z^m - Z^m_{id} \tag{1}$$

where Z^m is the thermodynamic function of mixing for a real system; Z^m_{id} is the thermodynamic function for an ideal mixture.

For binary mixtures, the Z^m_{id} values can be calculated using Equation 2:

$$Z^m_{id} = w_1 Z^0_1 + w_2 Z^0_2 \tag{2}$$

where Z^0_1 and Z^0_2 are the thermodynamic function values for pure water and pure protein, and w_1 and w_2 are the mass fractions of water and protein, respectively ($w_2 = 1 - w_1$).

For binary mixtures, the Z^E values are composed of two components:

$$Z^E = w_1 \overline{Z}^E_1 + w_2 \overline{Z}^E_2 \tag{3}$$

where $\overline{Z}^E_1$ is the excess function for component 1 (water), and $\overline{Z}^E_2$ is the excess function for component 2 (protein).

The excess partial quantities of components 1 and 2 are calculated using Equations 4 and 5:

$$\overline{Z}_1^E = Z^E - w_2(\frac{dZ^E}{dw_2})_{T,P} \tag{4}$$

$$\overline{Z}_2^E = Z^E - w_1(\frac{dZ^E}{dw_1})_{T,P} \tag{5}$$

where $\overline{Z}_1^E$ and $\overline{Z}_2^E$ are the excess partial quantities of components 1 and 2.

2.2. Excess Partial Volumes of Water and Proteins

The excess partial volumes of proteins and water are calculated using Equations 6 and 7:

$$\overline{V}_1^E = V^E - w_2(\frac{dV^E}{dw_2})_{T,P} \tag{6}$$

$$\overline{V}_2^E = V^E - w_1(\frac{dV^E}{dw_1})_{T,P} \tag{7}$$

where $\overline{V}_1^E$ and $\overline{V}_2^E$ are the excess partial volumes (cm^3 g^{-1}) of the water and protein.

2.3. Excess Partial Heat Capacities of Water and Proteins

The excess partial heat capacities at 25 °C and atmospheric pressure can be calculated using Equations 8 and 9:

$$\overline{C}_{P1}^E = C_P^E - w_2(\frac{dC_P^E}{dw_2})_{T,P} \tag{8}$$

$$\overline{C}^E_{P2} = C^E_P - w_1\left(\frac{dC^E_P}{dw_1}\right)_{T,P} \tag{9}$$

where $\overline{C}^E_{P1}$ and $\overline{C}^E_{P2}$ are the excess partial heat capacities of water (J g^{-1} water K^{-1}) and protein (J g^{-1} protein K^{-1}), respectively.

2.4. Excess Partial Enthalpies

The hydration of proteins can be characterized using two different methods.

2.4.1. Method 1

This method is typically used to characterize the hydration of proteins [1-3]. The total enthalpy change corresponding to the mixing of water with the protein, $\Delta_{tot} H$, is related to the amount of water bound to the protein, m^h_w, (Equation 10):

$$\Delta H_{hydr} = \frac{\Delta_{tot} H}{m^h_w} = [\overline{H^h_P} - \overline{H^0_P}]\frac{m_P}{m^h_w} + [\overline{H}^h_{w,.protein} - \overline{H}_{w,.liquid}] \tag{10}$$

where $\overline{H}^0_P$ is the partial enthalpy of the dried protein at 25°C and atmospheric pressure; $\overline{H}^h_P$ is the partial enthalpy of the hydrated protein; $\overline{H}^h_{w,.protein}$ is the partial enthalpy of water bound to the protein; $\overline{H}_{w,.liquid}$ is the partial enthalpy of water in pure liquid state at 25°C and atmospheric pressure; m_P is the amount of the protein.

The ΔH_{hydr} value is also referred to as the enthalpy of protein hydration. As concluded from Equation 10, *Method 1* does not facilitate the estimation of the individual contributions of the enthalpies of the protein and water to the calorimetrically measured heat effects. As shown in Equation 10, the ΔH_{hydr} value contains two components.

The first component, $\overline{[H_P^h} - \overline{H_P^0]}\frac{m_P}{m_w^h}$, reflects changes in the protein state during hydration (including conformational rearrangements and glass transition). The second component, $[\overline{H}^h_{w,.protein} - \overline{H}_{w,.liquid}]$, reflects the difference between the water state in the liquid phase and the bound state.

2.4.2. Method 2

This method facilitates the estimation of the individual protein and water partial quantities in the entire range of water content. The interaction enthalpy observed during the addition of water to the protein is considered to be an excess quantity of the mixing of water with the protein (H^E, Equation 3). The H^E value is calculated from the enthalpies of the solution of the dried and hydrated protein in pure liquid water. The enthalpy of the solution of the dried protein in pure liquid water, $\Delta_{sol}H(dried)$, is described in Equation 11. The final state of the protein (Equations 11 and 12) is a diluted solution in water at 25°C and at atmospheric pressure:

$$\Delta_{sol}H(dried) = [\overline{H}_P^{sol} m_P] - [\overline{H}_P^o m_P] + [\overline{H}^{sol}_{w,.protein} m_w^{sol}] - [\overline{H}_{w,.liquid} m_w^{sol}] \quad (11)$$

where $\overline{H}_P^0$ is the partial enthalpy of the dried protein at 25°C and at atmospheric pressure; $\overline{H}_{w,.liquid}$ is the partial enthalpy of water in pure liquid state at 25°C and at atmospheric pressure; $\overline{H}^{sol}_{w,.protein}$ is the partial enthalpy of water bound to the protein in the diluted aqueous solution; $\overline{H}_P^{sol}$ is the partial enthalpy of the protein in the solution; m_P is the amount of the protein; m_w^h is the amount of water bound to the protein; and m_w^{sol} is the amount of water bound to the protein in the diluted aqueous solution.

The enthalpy of solution of the initially-hydrated protein in pure liquid water, $\Delta_{sol}\mathrm{H}(hydrated)$, is described in Equation 12:

$$\Delta_{sol}\mathrm{H}(hydrated) = [\overline{\mathrm{H}}_P^{sol}\mathrm{m}_P] - [\overline{\mathrm{H}}_P^{h}\mathrm{m}_P] + [\overline{\mathrm{H}}_{w,.protein}^{sol}\mathrm{m}_w^{sol}] - [\overline{\mathrm{H}}_{w,.liquid}\mathrm{m}_w^{sol}] + [\overline{\mathrm{H}}_{w,.liquid}\mathrm{m}_w^{h}] - [\overline{\mathrm{H}}_{w,.protein}^{h}\mathrm{m}_w^{h}] \quad (12)$$

where $\overline{H}_{\mathrm{w,.protein}}^{\mathrm{h}}$ is the partial enthalpy of water bound to the protein, and $\overline{H}_{\mathrm{P}}^{\mathrm{h}}$ is the partial enthalpy of the hydrated protein. As shown in Equation 13, the H^{E} value is calculated from the enthalpies of solution of the dried, $\Delta_{sol}\mathrm{H}(dried)$, and hydrated, $\Delta_{sol}\mathrm{H}(hydrated)$, protein in water, measured using isothermal calorimetry:

$$\frac{\Delta_{sol}\mathrm{H}(dried) - \Delta_{sol}\mathrm{H}(hydrated)}{\mathrm{m}_w^h + \mathrm{m}_P} = [\overline{\mathrm{H}_P^h} - \overline{\mathrm{H}_P^0}]\frac{\mathrm{m}_P}{\mathrm{m}_P + \mathrm{m}_w^h} + [\overline{\mathrm{H}}_{w,.protein}^h - \overline{\mathrm{H}}_{w,.liquid}]\frac{\mathrm{m}_w^h}{\mathrm{m}_P + \mathrm{m}_w^h} = [\overline{\mathrm{H}_P^h} - \overline{\mathrm{H}_P^0}]\mathrm{w}_2 + [\overline{\mathrm{H}}_{w,.protein}^h - \overline{\mathrm{H}}_{w,.liquid}]\mathrm{w}_1 = \mathrm{H}^E \quad (13)$$

The excess partial enthalpies of the protein and water are calculated using Equations 14 and 15:

$$\overline{\mathrm{Z}}_1^E = \mathrm{Z}^E - \mathrm{w}_2(\frac{d\mathrm{Z}^E}{d\mathrm{w}_2})_{T,P} = [\overline{\mathrm{H}}_{w,.protein} - \overline{\mathrm{H}}_{w,.liquid}] = \overline{\mathrm{H}}_1^E \quad (14)$$

$$\overline{Z}_2^{\mathrm{E}} = Z^{\mathrm{E}} - w_1(\frac{dZ^{\mathrm{E}}}{dw_1})_{\mathrm{T,P}} = [\overline{H_{\mathrm{P}}^{\mathrm{h}}} - \overline{H_{\mathrm{P}}^{0}}] = \overline{H}_2^{\mathrm{E}} \quad (15)$$

where $\overline{H}_1^E$ (J g^{-1} water) and $\overline{H}_2^E$ (J g^{-1} protein) are the excess partial enthalpies of the water and protein, respectively, and w_1 and w_2 are the mass fractions of water and protein, respectively. The symbols used in Section 2.4 "Excess partial enthalpies" are summarized in Table 1.

Table 1. Symbols Used in Section 2.4. "Excess partial enthalpies"

Symbol	Dimension	Name
W	-	Water
P	-	Protein
h	-	Hydrated
m_W^h	g water	Amount of water bound to the protein
m_P	g protein	Amount of the protein
m_W^{sol}	g water	Amount of water bound to the protein in the diluted aqueous solution
W_1	-	Mass fraction of water
W_2	-	Mass fraction of protein
$\overline{H}_{w,.liquid}$	J g^{-1} water	Partial enthalpy of water in pure liquid state at 25°C and atmospheric pressure
$\overline{H}^h_{w,.protein}$	J g^{-1} water	Partial enthalpy of water bound to the protein at 25°C and atmospheric pressure
$\overline{H}^0_P$	J g^{-1} protein	Partial enthalpy of the dried protein at 25°C and atmospheric pressure
$\overline{H}^h_P$	J g^{-1} protein	Partial enthalpy of the hydrated protein at 25°C and atmospheric pressure
$\overline{H}^{sol}_{w,..protein}$	J g^{-1} water	Partial enthalpy of water bound to the protein in the diluted aqueous solution
$\overline{H}^{sol}_P$	J g^{-1} protein	Partial enthalpy of the protein in the diluted aqueous solution
$\Delta_{sol} H(dried)$	J	Enthalpy of solution of the dried protein in pure water at 25°C and atmospheric pressure
$\Delta_{sol} H(hydrated)$	J	Enthalpy of solution of the hydrated protein in pure water at 25°C and atmospheric pressure

2.5. Excess Partial Gibbs Energies

The Gibbs-Duhem equation ($\sum_i n_i d\mu_i = 0$) was applied to calculate the excess Gibbs free energies. For a binary mixture, the Gibbs-Duhem equation can be described using Equation 16:

$$n_1 d\mu_1 + n_2 d\mu_2 = 0 \tag{16}$$

where μ_1 and μ_2 are the partial Gibbs energies of water and the protein, respectively.

This implies that if one partial quantity increases, the other must decrease:

$$d\mu_2 = -\frac{n_1}{n_2} d\mu_1 = -\frac{x_1}{x_2} d\mu_1 \tag{17}$$

where x_1 and x_2 are the water and protein mole fractions in the water-protein mixtures, respectively.

At constant pressure and temperature:

$$d\mu_i = RTd(lna_i) \tag{18}$$

where a_i is the water or protein activity in the water-protein mixtures.

Equation 17 can be transformed into Equation 19:

$$d(lna_2) = -\frac{1-x_2}{x_2} d(lna_1) \tag{19}$$

The water and protein activity coefficients, γ_i, are calculated using Equation 20:

$$a_i = \gamma_i * x_i \tag{20}$$

where x_i is the water or protein mole fraction in the water-protein mixtures, and γ_i is the water or protein activity coefficient in the water-protein mixtures.

For non-ideal systems, Equation 19 can be transformed into Equation 21:

$$d(ln\gamma_2)+d(lnx_2)= =-\frac{1-x_2}{x_2}d(ln\gamma_1)-\frac{1-x_2}{x_2}d(ln(1-x_2)) \tag{21}$$

For ideal systems, Equation 19 can be transformed into Equation 22:

$$d(lnx_2)=-\frac{1-x_2}{x_2}d(ln(1-x_2)) \tag{22}$$

The difference between Equation 19 and Equation 22 is:

$$d(ln\,\gamma_2)=-\frac{1-x_2}{x_2}d(ln\,\gamma_1) \tag{23}$$

The integration of Equation 23 from $\ln(\gamma_1^0)$ at x_2=1 to $\ln(\gamma_1)$ at x_2^* gives Equation 24:

$$ln\,\gamma_2=-\int_{ln(\gamma_1^0)}^{ln(\gamma_1)}\frac{1-x_2}{x_2}d(ln\,\gamma_1)= =-\int_{ln(\gamma_1^0)}^{ln(\gamma_1)}\frac{n_1}{n_2}d(ln\,\gamma_1) \tag{24}$$

The excess partial Gibbs energy of water and protein can be calculated using Equation 25 and Equation 26:

$$\mu_2^E = -\int_{-\infty}^{\mu_1^E} \frac{n_1 * M_1}{n_2 * M_2} d(\mu_1^E) =$$

$$= -\int_{-\infty}^{\mu_1^E} \frac{w_1}{w_2} d(\mu_1^E) \tag{25}$$

$$\mu_1^E = \frac{RT}{M_1} ln\, \gamma_1 \tag{26}$$

where μ_1^E (J/g water) is the excess partial Gibbs energies of water; μ_2^E (J/g protein) is the excess partial Gibbs energy of protein; w_1 and w_2 are the weight fractions of water and protein, respectively; M_1 is the molecular mass of water; and M_2 is the molecular mass of lysozyme.

The excess Gibbs energy of mixing (J/g mixture) can be calculated using Equation 27:

$$G^E = w_1 \mu_1^E + w_2 \mu_2^E \tag{27}$$

2.6. Excess Partial Entropies

The excess partial entropies are calculated using Equations 28-30:

$$\frac{H^E - G^E}{T} = S^E \tag{28}$$

$$\overline{S}_1^E = S^E - w_2 (\frac{dS^E}{dw_2})_{P,T} \tag{29}$$

$$\overline{S}_2^E = S^E - w_1\left(\frac{dS^E}{dw_1}\right)_{P,T} \tag{30}$$

where $\overline{S}_1^E$ and $\overline{S}_2^E$ are the excess partial entropies of water and protein.

2.7. Isosteric Quantities

The isosteric quantity is the change in partial molar quantity for the transfer of water from the vapor phase to the protein at water equilibrium pressure P. These values are estimated using Equation 31. Equation 31 is applied for constant (isosteric) water content (Γ).

$$\ln P_\Gamma = \frac{\overline{\Delta S_\Gamma}}{R} - \frac{\overline{\Delta H_\Gamma}}{R}\left(\frac{1}{T}\right) = \frac{\overline{S}_V - \overline{S}_P}{R} - \frac{\overline{H}_V - \overline{H}_P}{R}\left(\frac{1}{T}\right) \tag{31}$$

where $\overline{S}_V$ is the partial molar entropy of water in the vapor phase; $\overline{S}_P$ is the partial molar entropy of the sorbed water; $\overline{H}_V$ is the partial molar enthalpy of water in the vapor phase; $\overline{H}_P$ is the partial molar enthalpy of the sorbed water; and R is the gas constant.

The isosteric quantities are estimated from the temperature dependence of the water sorption isotherm (The van't Hoff plots [lnP versus 1/T] at various lysozyme water content [for example, Figure 6 from Reference [21]).

References

[1] Gregory, R. B. (1995) Protein hydration and glass transition behavior. In: Gregory, R. B., editor. *Protein-Solvent Interactions*, Marcel Dekker, New York, 191-264.

[2] Rupley, J. A., Careri, G. (1991) Protein hydration and function. *Adv. Protein Chem.*, 41, 37-172.

[3] Kuntz, I. D., Kauzmann, W. (1974) Hydration of proteins and polypeptides. *Adv. Protein. Chem.*, 28, 239-345.

[4] Oleinikova, A., Smolin, N., Brovchenko, I., Geiger, A., Winter, R. (2005) Properties of spanning water networks at protein surfaces. *J. Phys. Chem. B*, 109, 1988-1998.

[5] Sirotkin, V. A., Solomonov, B. N., Faizullin, D. A., Fedotov, V. D. (2002) Sorption of water and acetonitrile by human serum albumin. *Russ. J. Phys. Chem. A*. 76: 2051-2057.

[6] Sirotkin, V. A., Winter, R. (2010) Volume changes associated with guanidine hydrochloride, temperature, and ethanol induced unfolding of lysozyme. *J. Phys. Chem. B*, 114, 16881-16886.

[7] Durchschlag, H., Zipper, P. (2001) Comparative investigations of biopolymer hydration by physicochemical and modeling techniques. *Biophys. Chem.*, 93, 141-.

[8] Kim, K., Kauzmann, W. (1980) Concentration dependence of the specific volume of human oxyhemoglobin, bovine serum albumin, ovalbumin solutions. *J. Phys. Chem.*, 84, 163-165.

[9] Bernhardt, J., Pauly, H. (1980) Pycnometric vs. densimetric measurements of highly viscous protein solutions. *J. Phys. Chem.*, 84, 158-162.

[10] Bull, B. H., Breese, K. (1968) Protein hydration. II. Specific heat of egg albumin. *Arch. Biochem. Biophys.*, 128, 497-502.

[11] Neurath, H., Bull, H. B. (1936) Denaturation and hydration of proteins. *J. Biol. Chem.,* 115, 519-528.

[12] Low, B. W., Richards, F. M. J. (1954) Measurements of the density, composition and related unit cell dimensions of some protein crystals. *J. Amer. Chem. Soc.*, 76, 2511-2518.

[13] McMeekin, T. L., Groves, M. L., Hipp, N. J. (1954) Partial specific volume of the protein and water in beta-lactoglobulin crystals, *J. Polymer. Sci.*, 12, 309-315. 157.

[14] Sirotkin, V. A., Khadiullina, A. V. (2011) Hydration of proteins: excess partial enthalpies of water and proteins. *J. Phys. Chem. B*, 115, 15110-15118.

[15] Sirotkin, V. A., Komissarov, I. A., Khadiullina, A. V. (2012) Hydration of proteins: excess partial volumes of water and proteins, *J. Phys. Chem. B*, 116, 4098-4105.

[16] Sirotkin, V. A., Khadiullina, A. V. (2014) A study of the hydration of ribonuclease A using densitometry: Effect of the protein hydrophobicity and polarity. *Chem. Phys. Lett.*, 603, 13-17.

[17] Sirotkin, V. A., Khadiullina, A. V. (2013) Gibbs energies, enthalpies, and entropies of water and lysozyme at the inner edge of excess hydration. *J. Chem. Phys.*, 139, 075102/1-075102/9.
[18] Atkins, P. W. (2006) *Physical Chemistry*. 8th ed. Oxford: Oxford University Press.
[19] Prausnitz, J. M. (1969) *Molecular Thermodynamics of Fluid-Phase Equilibria*. N.J.: Prentice-Hall, Inc., Engelwood Cliffs.
[20] Belousov, V. P., Panov, M. Y. (1994) *Thermodynamic properties of aqueous solutions of organic substances*. Boca Raton, Fla.: CRC Press.
[21] Bone, S. (1996) Dielectric and gravimetric studies of water binding to lysozyme. *Phys. Med. Biol.* 41, 1265-1275.

In: Protein – Water Interactions
Editor: Vladimir A. Sirotkin
ISBN: 978-1-63463-007-8
© 2014 Nova Science Publishers, Inc.

Chapter 2

A STUDY OF THE EXCESS GIBBS ENERGY OF RIBONUCLEASE A - WATER MIXTURES

Vladimir A. Sirotkin *** *and Aigul V. Khadiullina***
A. M. Butlerov Institute of Chemistry,
Kazan Federal University, Kazan, Russia

ABSTRACT

Water sorption measurements were applied to characterize the hydration dependencies of the excess thermodynamic functions of binary protein-water mixtures. The aim of this study is to demonstrate how these quantities correlate with the coverage of the protein macromolecules by water molecules. The advantages of our methodology are (i) we are able to simultaneously determine the excess partial quantities of water and proteins; (ii) these thermodynamic quantities can be determined in the entire range of water content. Here, in particular, the excess Gibbs energies of the binary system of bovine pancreatic ribonuclease A (RNase A) with water were obtained as a function of composition at 25 °C. The results from the thermodynamic measurements were analysed to give a unified picture of the hydration process of proteins.

Keywords: Gibbs energy, Protein (biomacromolecule) hydration, Excess function, Partial quantity

---

* Corresponding author: Vladimir A. Sirotkin. A. M. Butlerov Institute of Chemistry, Kazan Federal University, Kremlevskaya street, 18, Kazan, 420008, Russia. E-mail: vsir@mail.ru.

INTRODUCTION

The Gibbs energy of mixing water with a protein is related to the water activity, a_w. This quantity can easily be measured experimentally from water sorption isotherms. Sorption isotherms (water content as a function of water activity) of many proteins have been reliably measured [1-13]. As is typical of proteins, the sorption isotherm has a sigmoid shape.

The partial molar Gibbs energies of the mixing of water with lysozyme were estimated in the water weight fraction from 0 to 0.35 [6, 12]. It was shown that, in the studied range, the partial Gibbs energies of water were negative. They decreased with increasing water content and were close to zero at a water weight fraction (w_1) of 0.35.

Foss and Reyerson studied the water sorption by ribonuclease in the range of 0-8% (g g^{-1}) [13]. From the temperature dependence of the water sorption isotherms in the range 20-30 °C, they calculated differential enthalpies and entropies of water sorption. Luscher-Mattli and Ruegg [5, 6]. calculated the entropies of water sorption by lysozyme and α-chymotrypsin. They studied the water sorption isotherms in the water weight fraction range of 0 to ~0.4. The hydration entropies were calculated from the temperature dependence of the water vapor pressure in the range of 10-40 °C. Bone studied the water sorption by lysozyme in the range of 1.5-19% (g g^{-1}) [7]. Calculations were performed using the temperature dependence of the water vapor pressure in the range of 6-46 °C. Sorption calorimetry has been used to measure the absorption isotherm of water by lysozyme and the corresponding hydration entropies in a water weight fraction of 0 to 0.35 [12]. The most important observations of these calorimetric studies [1-13] can be summarized as follows: The hydration Gibbs energies (enthalpies, enthalpies) vary strongly with the hydration level. They are large at low hydration levels. As the water content increases, the hydration enthalpies approach the enthalpy of condensation of pure water.

Calorimetry is one of the most effective methods for obtaining reliable thermodynamic information about the interactions of proteins with water in various environments. Amberg measured the heat of water vapor sorption using bovine serum albumin within a water content range of 0-12% (g g^{-1}) at 20 °C [14]. The enthalpies of the solution of the dried and hydrated ribonuclease A (RNase A) were measured using isothermal calorimetry at 25 °C [15]. The water content of RNase A was varied from 0 to 26% (g g^{-1}), and the enthalpy of the solution was negative (exothermic). As the water content of RNase A increases, the enthalpy of the solution approaches zero.

Smith et al. calorimetrically measured the heat of water adsorption by lysozyme in the range of relative water vapor pressures from 0 to 0.895 and obtained both the sorption isotherm and the enthalpy of hydration of the protein within a water content range of 0-18% (g g^{-1}) at 25 °C [16]. Sorption calorimetry has been used to measure the adsorption isotherm of water by lysozyme and the corresponding heat effects in the entire range of water activities [17]. The interaction enthalpy depends significantly on the initial water content and hydration history [14-17].

The obtained water sorption isotherms, Gibbs energies, enthalpies, and entropies [1-17] contain complete information about the binary water-protein systems, including the corresponding conformational changes in the protein structure and glass transition. However, the simultaneous estimation of the protein and water contributions to the Gibbs energy, enthalpy, and entropy of binary protein-water systems in the entire range of water content has not yet been attempted. A novel methodology has recently been proposed for simultaneously estimating the excess partial quantities of water and proteins [18-21]. This method is based on the analysis of the thermodynamic functions of mixing. The aim of this study was to determine the excess partial Gibbs energies of water and protein, and demonstrate how these quantities correlate with the coverage of the protein macromolecules by water molecules.

Bovine pancreatic ribonuclease A (RNase A) was used as a model. This biomolecule is one of the most hydrophilic proteins applied in biothermodynamic studies [22-24]. Ribonuclease A is a small monomeric enzyme of 124 amino acids, containing no non-enzyme components. Ribonuclease is an example of an $\alpha+\beta$ protein. RNase hydrolyzes single-stranded RNA through cleavage of the phosphodiester linkage at the 3' side of pyrimidine nucleotides [22-24].

2. Methodology

2.1. Excess Partial Quantities

The thermodynamic properties of a real binary water-protein system can be expressed in terms of the excess functions, Z^E, i.e., the difference between the observed thermodynamic function of mixing, Z^m, and the function for an ideal binary mixture, Z^m_{id}:

$$Z^E = Z^m - Z^m_{id} \tag{1}$$

The Z^m_{id} values can be calculated using Equation 2:

$$Z^m_{id} = w_1 Z^0_1 + w_2 Z^0_2 \tag{2}$$

where Z^0_1 and Z^0_2 are the thermodynamic function values for pure water and pure protein, and w_1 and w_2 are the mass fractions of water and protein, respectively.

Deviations of the excess functions from zero indicate the extent to which the studied binary system is non-ideal due to strong specific interactions between components (i.e., hydrogen bonding and charge-charge interactions).

The Z^E values are composed of two components:

$$Z^E = w_1 \overline{Z}^E_1 + w_2 \overline{Z}^E_2 \tag{3}$$

where $\overline{Z}^E_1$ is the excess partial function for component 1 (water), and $\overline{Z}^E_2$ is the excess partial function for component 2 (protein).

2.2. Excess Partial Gibbs Energies

The Gibbs-Duhem equation ($\sum_i n_i d\mu_i = 0$) was applied to calculate the excess Gibbs free energies. For a binary mixture, the Gibbs-Duhem equation can be described using Equation 4:

$$n_1 d\mu_1 + n_2 d\mu_2 = 0 \tag{4}$$

where μ_1 and μ_2 are the partial Gibbs energies of water and the protein, respectively.

This implies that if one partial quantity increases, the other must decrease:

$$d\mu_2 = -\frac{n_1}{n_2}d\mu_1 = -\frac{x_1}{x_2}d\mu_1 \quad (5)$$

where x_1 and x_2 are the water and protein mole fractions in the water-protein mixtures, respectively.

At constant pressure and temperature:

$$d\mu_1 = RTd(lna_i) \quad (6)$$

where a_i is the water or protein activity in the water-protein mixtures.

Equation 5 can be transformed into Equation 7:

$$d(lna_2) = -\frac{1-x_2}{x_2}d(lna_1) \quad (7)$$

The water and protein activity coefficients, γ_i, are calculated using Equation 8:

$$a_i = \gamma_i * x_i \quad (8)$$

where x_i is the water or protein mole fraction in the water-protein mixtures, and γ_i is the water or protein activity coefficient in the water-protein mixtures.

For non-ideal systems, Equation 7 can be transformed into Equation 9:

$$d(ln\gamma_2) + d(lnx_2) = \\ = -\frac{1-x_2}{x_2}d(ln\gamma_1) - \frac{1-x_2}{x_2}d(ln(1-x_2)) \quad (9)$$

For ideal systems, Equation 7 can be transformed into Equation 10:

$$d(\ln x_2) = -\frac{1-x_2}{x_2} d(\ln(1-x_2)) \tag{10}$$

The difference between Equation 7 and Equation 10 is:

$$d(\ln \gamma_2) = -\frac{1-x_2}{x_2} d(\ln \gamma_1) \tag{11}$$

The integration of Equation 11 from $\ln(\gamma_1^0)$ at x_2=1 to $\ln(\gamma_1)$ at x_2^* gives Equation 12:

$$\ln \gamma_2 = -\int_{\ln(\gamma_1^0)}^{\ln(\gamma_1)} \frac{1-x_2}{x_2} d(\ln \gamma_1) =$$

$$= -\int_{\ln(\gamma_1^0)}^{\ln(\gamma_1)} \frac{n_1}{n_2} d(\ln \gamma_1) \tag{12}$$

The excess partial Gibbs energy of water and protein can be calculated using Equation 13 and Equation14:

$$\mu_2^E = -\int_{-\infty}^{\mu_1^E} \frac{n_1 * M_1}{n_2 * M_2} d(\mu_1^E) =$$

$$= -\int_{-\infty}^{\mu_1^E} \frac{w_1}{w_2} d(\mu_1^E) \tag{13}$$

$$\mu_1^E = \frac{RT}{M_1} \ln \gamma_1 \tag{14}$$

where μ_1^E (J/g water) is the excess partial Gibbs energies of water; μ_2^E (J/g protein) is the excess partial Gibbs energy of protein; w_1 and w_2 are the weight fractions of water and protein, respectively; M_1 is the molecular mass of water; and M_2 is the molecular mass of lysozyme.

The excess Gibbs energy of mixing (J/g mixture) can be calculated using Equation 15:

$$G^E = w_1\mu_1^E + w_2\mu_2^E \tag{15}$$

3. Experimental Section

3.1. Materials

Bovine pancreatic ribonuclease A (RNase A) of the highest commercially-available purity was purchased from Sigma Chemical (St. Louis, MO) and used without further purification. The purity of the protein samples was verified through electrophoresis and dynamic light-scattering measurements (90Plus Particle Size Analyzer, Brookhaven Instruments Corporation, US) to be more than 95%. The molecular weight of the protein was taken as 13,700 Da. The water used was doubly-distilled.

3.2. Water Sorption Measurements

The dried protein preparation (zero hydration level) was obtained through vacuum drying using a microthermoanalyzer “Setaram” MGDTD-17S (±0.00001 g) at 25 °C and 0.1 Pa, until a constant sample weight was reached. The water content of the dried protein was estimated as 0.003 ± 0.002 g water g^{-1} protein using the Karl Fischer titration method, according to the recommendations [25, 26]. At the lowest and intermediate water weight fractions, the protein preparations were prepared as follows. The initially dehydrated protein samples were exposed to pure water vapor. The water vapor was consecutively flowed through a thermostated glass tube with drying agent (P_2O_5), a thermostated saturator filled with saturated salt solution, and a cell containing the protein sample. The schematic representation of the experimental setup is given in Figure 1 from Reference [21]. An external ethylene glycol thermostat (RC 6 from Lauda, Germany) was used to control the temperature with a precision of 0.1 °C. Sorption equilibrium was reached after 180 min. The water activity (a_w) in the vapor phase was adjusted by changing the saturated salt solution in the saturator. The water activities of the saturated salt solutions were taken from [27, 28]. The following salts were used: LiBr (a_w=0.064), KOH (a_w=0.078), LiCl (a_w=0.11), $CaBr_2$ (a_w=0.17), CH_3COOK (a_w=0.22), $MgCl_2$ (a_w=0.33), K_2CO_3 (a_w=0.44), $Mg(NO_3)_2$ (a_w=0.53), NaCl (a_w=0.75), KCl (a_w=0.84), KNO_3 (a_w=0.94), and K_2SO_4 (a_w=0.97). The salts used for sample conditioning were of analytical pure grade. The water content of the samples after equilibration was measured by drying under vacuum using a microthermoanalyzer "Setaram" MGDTD-17S at 25 °C and 0.1 Pa, until a constant sample weight was reached.

At the highest water weight fractions (a_w>0.98), the water activity was adjusted by changing the unsaturated NaCl-water solutions at 25 °C. The a_w values of the NaCl-water solutions were taken from Reference [29]. This technique was used for the protein-water systems for $w_1 > 0.6$.

4. Results and Discussion

Figure 1 shows the dependency of the μ_1^E values on the hydration level of RNase A. The μ_1^E values were calculated using Equation 14.

Figure 2 presents the μ_2^E values as a function of the weight mass of water. The μ_2^E values were calculated using Equation 13.

Figure 3 shows the dependency of the G^E values (excess function of mixing per-unit-mass of the water-protein mixture) on the hydration level of RNase A. The G^E values were estimated using Equation 15.

In ideal binary mixtures (mixtures of two components, W [water] and P [protein]) the average W-P interactions are the same as the average W-W and P-P interactions. Non-ideal mixtures are composed of particles for which the W-W, P-P, and W-P interactions are all different.

As shown in Figure 3, the G^E values differ significantly from zero, indicating that the water-protein mixtures are non-ideal in the entire range of water content. Figure 3 present the dependencies of the $w_i\overline{Z}_i^E$ ($w_1\mu_1^E$, $w_2\mu_2^E$) functions plotted versus the water mass fraction.

These functions were calculated using Equation 3. They show the individual protein and water contributions to the excess thermodynamic functions in the entire range of water content. Figure 3 demonstrate that the protein contributions ($w_2\overline{Z}_2^E$) reach maximal values at w_1~0.25. However, the $w_1\overline{Z}_1^E$ functions reach maximal values at low water content (w_1~0.05).

The partial quantity curves presented in Figures 1-3 can be divided into four parts.

4.1. Part 1. (W_1 = 0-0.05)

The μ_2^E values are close to zero and do not depend on the water content (Figure 2). At the lowest water content, the proteins are in a glassy (rigid) state [1]. No biological activity was observed at the lowest water content (Figure 3 from Reference [18]).

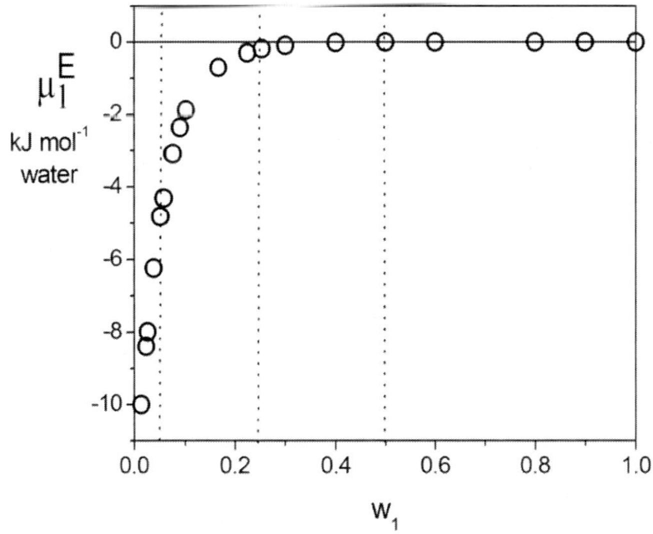

Figure 1. Ribonuclease A-water mixtures. Excess Gibbs energy of water, μ_1^E, as a function of the mass fraction of water, w_1, at 25°C Standard errors of estimation of the μ_1^E values were 0.2-0.3 kJ mol^{-1} water.

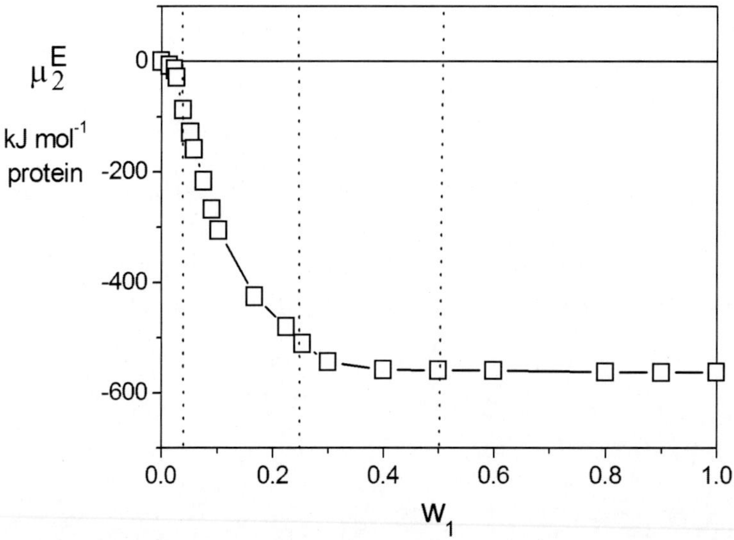

Figure 2. Ribonuclease A-water mixtures. Excess Gibbs energy of RNase A, μ_2^E, as a function of the mass fraction of water, w_1, at 25°C.

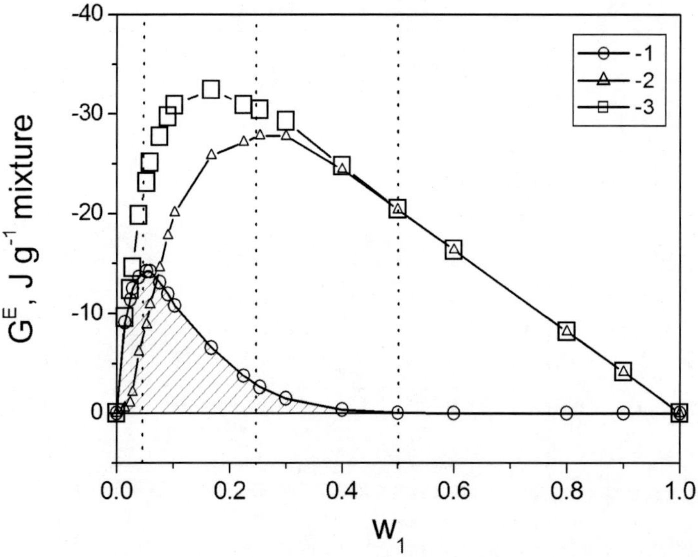

Figure 3. Ribonuclease A-water mixtures. Excess thermodynamic functions plotted versus the mass fraction of water, w_1, at 25 °C: 1 - $w_1\mu_1^E$; 2 - $w_2\mu_2^E$; 3 - G^E.

The fact that the dehydrated proteins are in the glassy state explains this feature of part 1. The lysozyme excess partial quantities are close to zero, reflecting the fact that all of the protein molecules contact the same protein molecules during this range of water content.

At $w_1<0.05$, the $w_2\overline{Z}_2^E$ values are close to zero (Figure 3). At $w_1\sim0.05$, the $w_1\overline{Z}_1^E$ functions reach maximal values.

These facts imply that, at the lowest water content, changes in the excess thermodynamic functions solely reflect water addition.

This region corresponds to the hydration of charged groups [1, 2, 45]. At low water content, the μ_1^E (excess partial Gibbs energy of water - Figure 1 [this study] and Figure 2 from Reference [21]), $\overline{TS}_1^E$ (excess partial entropy of water - Figure 5 from Reference [21]), and $\overline{H}_1^E$ (excess partial enthalpy of water - Figure 2 from Reference [18]) values differ significantly from zero. The μ_1^E, $\overline{H}_1^E$, and TS_1^E values are highly negative. As concluded in Reference [18], the $\overline{H}_1^E$ ($w_1=0$) values are close to those observed for diethylamine, suggesting that, at the lowest w_1 values, the interaction with the protein elements containing the amino groups might be a dominant factor controlling the state of water molecules.

4.2. Part 2. (W_1 = 0.05-0.25)

Proteins undergo a glasslike dynamic transition at water content of approximately 10 wt% at 25 °C [1]. This water content falls within part 2 in the present study. The transition from the glassy (rigid) to the flexible (elastic) state is accompanied by significant changes in the properties of the proteins [1]. For example, during the isothermal sorption of water, a glasslike transition results in a step in the μ_2^E (excess partial Gibbs energy of protein - Figure 1 [this study] and Figure 3 from Reference [21]), $\overline{TS}_2^E$ (excess partial entropy of protein - Figure 6 from Reference [21]), and $\overline{H}_2^E$ (excess partial enthalpy of protein - Figure 3 from Reference [18]) curves. The μ_2^E, $\overline{TS}_2^E$, and $\overline{H}_2^E$ values ([this study] and References [18, 21]) change sharply from extremely low values to highly negative values. The apparent heat capacity of lysozyme determined from isothermal experiments using a drop calorimeter, increases from extremely low values to high values in this water content interval [2, 31].

Part 2 corresponds to the hydration of polar groups [1, 2, 30]. In the water content range w_1 = *0.05-0.25,* the $\overline{TS}_1^E$ values change sharply from highly negative (~-9.0 kJ mol^{-1} water) to moderate (~-2.0 kJ mol^{-1} water) values [21]. This sharp transition reflects the formation of a spanning hydrogen-bonded network of water [2, 32]. The formation of this network occurs via a quasi two-dimensional percolation transition of the hydration water [32].

Quasielastic neutron and light-scattering techniques were employed to characterize the effect of hydration on the internal dynamics of lysozyme in the picoseconds-to-nanoseconds time range [33]. The increased hydration activates the fast relaxation process within part 2. Notably, the $\mu_1^E - w_1$, $\overline{TS}_1^E - w_1$, and $\overline{H}_1^E - w_1$ curves [this study, 18, 21], and the dependence of fast conformational fluctuations on the hydration level, are similar to the hydration dependence of hydrogen isotope exchange [2, 34].

4.3. Part 3. (W_1 = 0.25-0.5)

Part 3 indicates the appearance of the "structured" water [30]. "Structured" water is composed of molecules that interact with hydrophobic surface patches on the protein-bridging between bound water molecules. Rupley and Careri [2] asserted that this region contains condensed water molecules over weakly-interacting surface elements, which are likely nonpolar atoms, not polar groups or adjacent-charged.

The $\overline{TS}_1^E$ and $\overline{H}_1^E$ values alter, to a moderate degree, in this region of water content [18, 21]. The μ_1^E values vary between ~-1.0 and 0 kJ mol^{-1} water Figure 1 [this study] and [21].

Significant changes in the amide I and III regions of the dehydrated proteins were observed using Fourier transform infrared (FTIR) spectroscopy [35-37]. Klibanov and Griebenow examined several proteins and showed that dehydration lowers the α-helix content and increases the β-sheet content [37]. It was demonstrated that the β-sheet content in the dehydrated lysozyme state was approximately 44%; in the aqueous solution, the β-sheet content was about 18% [37].

The onset of biological activity was observed in this region. Rupley and Careri [2] examined the lysozyme-catalyzed hydrolysis of the hexasaccharide of N-acetylglucosamine [$(GlcNAc)_6$] (Figure 5 from Reference [19]) as a function of water content and showed that the reaction grows sharply at $w_1 > 0.25$. The μ_2^E, $\overline{TS}_2^E$, $\overline{H}_2^E$ values ([this study] and [18, 21]) and enzymatic activity ([2] and Figure 5 from Reference [19]) have different levels of dependency on hydration.

Enzymatic activity is suppressed within parts 1 and 2. Changes in the μ_2^E, $\overline{TS}_2^E$, and $\overline{H}_2^E$ values start at $w_1>0.05$ and are complete within part 3 ([this study] and [18, 21]).

The slow relaxation process [33] was activated at $w_1 > 0.25$ (Figure 5 from Reference [19]). It has been suggested that the slow relaxation process might be correlated with the motions of the secondary protein structures. The dependence of the slow process on the water content is associated with the hydration dependency of the enzymatic activity of lysozyme and the rotation relaxation time of a probe located near the protein surface, in studies performed with electron paramagnetic resonance (EPR) spectroscopy [2] (Figure 4 from Reference [18]). The slow relaxation process [33] is initiated, and the distribution of conformational states [1, 2] is close to normal, only when the hydration of the protein surface attains a particular level. When the formation of the spanning hydrogen bond network of water is complete, the proteins then become catalytically active. The correlations presented in Figures 1-3 show that the contributions corresponding to the hydrophilic and hydrophobic hydration of the proteins are coupled differently to the excess functions. The changes in the excess partial Gibbs energies, entropies and enthalpies corresponding to the hydration and polar and ionizable protein groups (hydrophilic hydration) are largely complete within regime 2. The changes in the excess partial quantities corresponding to the the hydration of nonpolar protein regions (hydrophobic hydration) are complete within regime 3. Thus, it is probable that regime 3 corresponds to the completion of the formation of the spanning hydrogen-bonded network of water.

4.4. Part 4. ($W_1 > 0.5$)

The high water content region is the biologically-important regime. Aggregation/association (protein-protein interactions) in the solutions at high water content is a topic of extensive research [38-41]. One of the possible reasons for this process is the disruption of the crucial hydration shell in the presence of low molecular additives (e.g., organic solvents and salts).

At the highest water content, the proteins are in a flexible (elastic) state [1]. Excess partial quantities attain their fully-hydrated values. The μ_2^E, $\overline{TS}_2^E$, and $\overline{H}_2^E$ values ([this study] and [18, 21]) reach minimal values. At $w_1<0.5$, the $w_1\overline{Z}_1^E$ values are near zero (Figure 3).

Bulk water was observed in this region through proton NMR measurements [30]. The molecular motion of these water molecules is solely determined through the interaction characteristics of water molecules. Water addition has no significant effect on the excess thermodynamic functions. At $w_1 > 0.5$, the changes in the excess thermodynamic functions primarily reflect changes in the state of ribonuclease A.

REFERENCES

[1] Gregory, R. B. (1995) Protein hydration and glass transition behavior. In: *Protein-Solvent Interactions* (Gregory, R. B., ed.), Marcel Dekker, New York, 191-264.

[2] Rupley, J. A., Careri, G. (1991) Protein hydration and function. *Adv. Protein Chem.,* 41, 37-172.

[3] Kuntz, I. D., Kauzmann, W. (1974) Hydration of proteins and polypeptides. *Adv. Protein Chem.,* 28, 239-345.

[4] Bull, H. B. (1944) Adsorption of water vapor by proteins. *J. Amer. Chem. Soc.* 66, 1499-1507.

[5] Luscher-Mattli, M., Ruegg, M. (1982) Thermodynamic functions of biopolymer hydration. I. Their determination by vapor pressure studies, discussed in an analysis of the primary hydration process. *Biopolymers* 21, 403-418.

[6] Luscher-Mattli, M., Ruegg, M. (1982) Thermodynamic functions of biopolymer hydration. II. Enthalpy-entropy compensation in hydrophilic hydration process. *Biopolymers* 21, 419-429.

[7] Bone, S. (1996) Dielectric and gravimetric studies of water binding to lysozyme. *Phys. Med. Biol.* 41, 1265-1275.

[8] Hnojewyj, W. S., Reyerson, L. H. (1961) Further studies on the sorption of H_2O and D_2O vapors by lysozyme and the deuterium-hydrogen exchange effect. *J. Phys. Chem.* 65, 1694-1698.

[9] Sirotkin V. A., Korolev, D. V. (2005) Effect of acetonitrile on the hydration of human serum albumin films: a calorimetric and spectroscopic study. *Thermochim. Acta,* 432, 246-253.

[10] Sirotkin, V. A., Solomonov, B. N., Faizullin, D. A., Fedotov, V. D. (2002) Sorption of water and acetonitrile by human serum albumin. *Russ. J. Phys. Chem. A,* 76, 2051-2057.

[11] Sirotkin, V. A. (2005) Effect of dioxane on the structure and hydration-dehydration of α-chymotrypsin as measured by FTIR spectroscopy. *Biochim. Biophys. Acta.*, 1750, 17-29.

[12] Kocherbitov, V., Arnebrant, T. (2010) Hydration of Lysozyme: The Protein-Protein Interface and the Enthalpy-Entropy Compensation. *Langmuir,* 26, 3918-3922.

[13] Foss, J. G., Reyerson, L. H. (1958) The sorption of water vapor by lyophilized ribonuclease. *J. Phys. Chem.* 62, 1214-1216.

[14] Amberg, C. H. (1957) Heats of adsorption of water vapor on bovine serum albumin. *J. Amer. Chem. Soc.,* 79, 3980-3984.

[15] Almog, R., Schrier, E. E. (1978) The dependence of the heat of solution of lyophilized solid ribonuclease A on water content. *J. Phys. Chem.,* 82, 1701-1702.

[16] Smith, A. L., Shirazi, H. M., Mulligan, S. R. (2002) Water sorption isotherms and enthalpies by lysozyme using the quartz crystal microbalance/heat conduction calorimeter. *Biochim. Biophys. Acta,* 1594, 150-159.

[17] Kocherbitov, V., Arnebrant, T., Söderman, O. (2004) Lysozyme-water interactions studied by sorption calorimetry. *J. Phys. Chem. B,* 108, 19036-19042.

[18] Sirotkin, V. A., Khadiullina, A. V. (2011) Hydration of proteins: excess partial enthalpies of water and proteins. *J. Phys. Chem. B*, 115, 15110-15118.

[19] Sirotkin, V. A., Komissarov, I. A., Khadiullina, A. V. (2012) Hydration of proteins: excess partial volumes of water and proteins, *J. Phys. Chem. B*, 116, 4098-4105.

[20] Sirotkin, V. A., Khadiullina, A. V. (2014) A study of the hydration of ribonuclease A using densitometry: Effect of the protein hydrophobicity and polarity. *Chem. Phys. Lett.*, 603, 13-17.

[21] Sirotkin, V. A., Khadiullina, A. V. (2013) Gibbs energies, enthalpies, and entropies of water and lysozyme at the inner edge of excess hydration. *J. Chem. Phys.*, 139, 075102/1-075102/9.

[22] Fersht, A. *Structure and mechanism in protein science: a guide to enzyme catalysis and protein folding*. New York: Freeman and Co, 1999.

[23] Lehninger, A. L., Nelson, D. L., Cox, M. M. *Principles of Biochemistry*. New York: Worth, 1993.

[24] Raines, R. T. (1998) Ribonuclease A. *Chem. Rev.* 98: 1045-1065.

[25] Borisover, M. D., Sirotkin, V. A., Solomonov, B. N. (1995) Thermodynamics of water binding by human serum albumin suspended in acetonitrile. *Thermochim. Acta*, 254, 47-53.

[26] Sirotkin, V. A., Zinatullin, A. N., Solomonov, B. N., Faizullin, D. A., Fedotov, V. D. (2002) Interaction enthalpies of solid bovine pancreatic chymotrypsin with organic solvents: comparison with FTIR-spectroscopic data. *Thermochim. Acta*, 382, 151-160.

[27] Nikolskii, B. P. (1963) *Chemist's Handbook,* Leningrad, Russia: Goskhimizdat.

[28] Greenspan, L. (1977) Humidity fixed points of binary saturated aqueous solutions. *J. Res. Natl. Bur. Stand.*, Sect. A 81, 89.

[29] Mazurkiewicz, J., Tomasik, P., Zaplotny, J. (2001) Relationships between water activity and viscosity of solutions. *J. Food Hydrocoll.* 15, 43-46.

[30] Fullerton, G. D., Ord, V. A., Cameron, I. L. (1986) An evaluation of the hydration of lysozyme by an NMR titration method. *Biochem. Biophys. Acta*, 869, 230-246.

[31] Yang, P.-H., Rupley, J. A. (1979) Protein-water interactions. Heat capacity of the lysozyme-water system. *Biochemistry*, 18, 2654-2661.

[32] Oleinikova, A., Smolin, N., Brovchenko, I., Geiger, A., Winter, R. (2005) Properties of spanning water networks at protein surfaces. *J. Phys. Chem. B*, 109, 1988-1998.

[33] Roh, J. H., Curtis, J. E., Azzam, S., Novikov, V. N., Peral, I., Chowdhuri, Z., Gregory, R., Sokolov, A. P. (2006) Influence of hydration on the dynamics of lysozyme, *Biophys. J.* 91, 2573-2588.

[34] Schinkel, J. E., Downer, N. W., Rupley, J. A. (1985) Hydrogen exchange of lysozyme powders. Hydration dependence of internal motions. *Biochemistry* 24, 352-358.

[35] Prestrelsky, S. J., Tedeschi, N., Arakawa, T., Carpenter, J. F. (1993) Dehydration-induced conformational transitions in proteins and their inhibition by stabilizers. *Biophys. J.,* 65, 661-671.

[36] Van de Weert, M., Harvez, P. I., Hennink, W. E., Crommelin, D. J. A. (2001) Fourier transform infrared spectrometric analysis of protein conformation: effect of sampling and stress factors. *Anal. Biochem.*, 297, 160-169.

[37] Griebenow, K., Klibanov, A. M. (1995) Lyophilization-induced reversible changes in the secondary structure of proteins. *Proc. Natl. Acad. Sci. US* 92, 10969-10976.

[38] Timasheff, S. N. (2002) Protein-solvent preferential interactions, protein hydration, and the modulation of biochemical reactions by solvent components, *Proc. Natl. Acad. Sci. US* 99, 9721-9726.

[39] Timasheff, S. N. (1992) Water as ligand: Preferential binding and exclusion of denaturants in protein unfolding, *Biochemistry* 31, 9857-9864.

[40] Kamiyama, T., Matsusita, T., Kimura, T. (2003) Volumetric study of lysozyme in dimethyl sulfoxide + water solution at 298.15 K, *J. Chem. Eng. Data* 48, 1301-1305.

[41] Sirotkin, V. A., Winter, R. (2010) Volume changes associated with guanidine hydrochloride, temperature, and ethanol induced unfolding of lysozyme. *J. Phys. Chem. B*, 114, 16881-16886.

In: Protein – Water Interactions
Editor: Vladimir A. Sirotkin
ISBN: 978-1-63463-007-8
© 2014 Nova Science Publishers, Inc.

Chapter 3

ANALYSIS OF HYDRATION OF OVALBUMIN BY DENSITOMETRY

Vladimir A. Sirotkin*[*] *and Aigul V. Khadiullina
Kazan Federal University, A. M. Butlerov
Institute of Chemistry, Kazan, Russia

ABSTRACT

High-precision densitometry was applied to study the hydration of hen egg-white ovalbumin. The excess volumes of the binary system of ovalbumin with water were obtained as a function of composition at 25 °C. The hydration process was characterised by analysing the excess functions of mixing. This method facilitates the individual evaluation of the protein and water partial quantities in the entire range of water content. The excess partial volumes are extremely sensitive to changes in the state of water and the protein. The excess volumes are determined by the hydration of the hydrophilic and hydrophobic protein groups. It was found that the more hydrophilic a protein is, the more significant the hydrophilic hydration contribution is.

Keywords: Protein (biomacromolecule) hydration, Densitometry, Volume, Excess functions

[*] Corresponding author: Vladimir A. Sirotkin. Kazan Federal University, A. M. Butlerov Institute of Chemistry, Kremlevskaya street, 18, Kazan, 420008, Russia. E-mail: vsir@mail.ru.

1. Introduction

Water plays a unique and ubiquitous role in the protein-water systems. Folding, stability, and functions of protein molecules are all strongly influenced by their interactions with water molecules [1-9]. Water can act as a plasticizer of protein conformation. Dehydrated proteins are rigid and glassy. Proteins undergo a glasslike transition at 25 °C and water content of about 10% (w/w) [1]. The transition from the glassy (rigid) to the flexible (elastic) state is accompanied by significant changes in the thermodynamic and structural properties. As the protein crosses the glass transition region into the flexible state, segmental motions and conformational rearrangements become possible and thermal expansivity is greatly increased. On the other hand, the protein–water interactions mediate the properties of hydration water surrounding the protein [1-9].

Volume is an important thermodynamic quantity directly associated to the compactness or globularity of a protein, and is generally dependent on a combination of factors [6, 10-13]. The hydration of charged and polar groups reduces volume. However, the volume changes associated with the exposure of hydrophobic groups depend on the model compounds selected, ranging from small negative to positive values. Moreover, it is not clear whether the volume change associated with hydrophobic hydration plays an important role in the total volume change.

Several studies have investigated the water density (volume) in the hydration shell of the proteins. A combined neutron and X-ray scattering study [14] showed that the average density of the first hydration shell of proteins is significantly higher than that of bulk water. This finding is consistent with the results obtained from the molecular simulation, crystallographic, and solution studies [10, 15-17]. The higher density of preferentially bound water and a comparison of hydration values in solution are also given in Reference [6].

The partial specific volumes for the majority of globular proteins in aqueous solutions fall within a narrow range between 0.70 and 0.75 $cm^3\ g^{-1}$ [3, 6, 10-13, 18, 19]. The concentration dependences of the apparent volumes of serum albumin, ovalbumin, and oxyhemoglobin were measured at high protein concentrations (protein mass fraction, w_2, ~0.3-0.4) [18].

No effect of protein on the solvent was observed at these concentrations, suggesting a constant partial specific volume of the solvent in the studied concentration range. The dilatometric measurements of serum albumin and oxyhemoglobin are consistent with this conclusion [19].

Direct volumetric studies of binary water-protein systems at low hydration levels are relatively rare. Volumes of protein-water systems with a w_1 (water mass fraction) between 0 and 0.56 were measured in References [20-23]. It was shown that at a w_1>0.2-0.3, the volumes of the binary protein-water systems depend linearly on the water content.

However, at the lowest w_1 values, there are deviations from this linearity. Thus, these studies [20-23] suggest that the partial specific volumes of proteins are lower at high hydration levels than in the dried state. Bull and Breese [20] estimated that at a w_1<0.2, the partial specific volume of water is lower than that of bulk water.

The measured volumes [10-13, 18-23] contain total information on the binary water-protein systems, including the corresponding conformational changes in the protein structure and the glass transition. However, the simultaneous estimation of the protein and water contributions to the volume of binary protein-water systems in the entire range of water content has not been attempted.

In this study, a novel methodology [8, 24] was applied to investigate the protein-water interactions. This methodology is based on the analysis of the excess thermodynamic functions of mixing. It facilitates the individual evaluation of the protein and water partial quantities in the entire range of water content.

The present study is aimed at understanding which molecular parameters control the excess volumes of binary protein-water systems.

To test the predictive ability of the proposed method, a system containing, as a model protein, hen egg-white ovalbumin was studied in this work. The excess partial quantities for ovalbumin were compared with the published data for several unrelated globular proteins (lysozyme, serum albumin, lactoglobulin) [8, 9]. These biomacromolecules represent a series of proteins in which the hydrophobicity of proteins is gradually changed over a wide range. Ovalbumin is the main protein found in egg white [25-27]. This protein consists of 385 amino acids residues. It is generally assumed to be a storage protein.

2. METHODOLOGY

The thermodynamic properties of a real binary water-protein system can be expressed in terms of the excess functions, Z^E. They are the difference between the thermodynamic function of mixing in a real system and the value corresponding to an ideal system at the same temperature, pressure and composition. For an ideal system, all excess functions are zero. Deviations of the excess functions from zero indicate the extent to which the studied binary system is non-ideal due to strong specific interactions between components (i.e., hydrogen bonding and charge-charge interactions). More comprehensive reviews of excess functions are given in References [28-30].

The Z^E values can be calculated using Equation 1:

$$Z^E = Z^m - Z^m_{id} \tag{1}$$

where Z^m is the thermodynamic function of mixing for a real system; Z^m_{id} is the thermodynamic function for an ideal mixture.

For binary mixtures, the Z^m_{id} values can be calculated using Equation 2:

$$Z^m_{id} = w_1 Z^0_1 + w_2 Z^0_2 \tag{2}$$

where Z^0_1 and Z^0_2 are the thermodynamic function values for pure water and pure protein, and w_1 and w_2 are the mass fractions of water and protein, respectively ($w_2 = 1 - w_1$).

For binary mixtures, the Z^E values are composed of two components:

$$Z^E = w_1 \overline{Z}^E_1 + w_2 \overline{Z}^E_2 \tag{3}$$

where $\overline{Z}^E_1$ is the excess function for component 1 (water), and $\overline{Z}^E_2$ is the excess function for component 2 (protein).

The excess partial quantities of components 1 and 2 are calculated using Equations 4 and 5:

$$\overline{Z}^E_1 = Z^E - w_2 \left(\frac{dZ^E}{dw_2}\right)_{T,P} \tag{4}$$

$$\overline{Z}_2^E = Z^E - w_1(\frac{dZ^E}{dw_1})_{T,P} \tag{5}$$

where $\overline{Z}_1^E$ and $\overline{Z}_2^E$ are the excess partial quantities of components 1 and 2.

The excess partial volumes of ovalbumin and water are calculated using Equations 6 and 7:

$$\overline{V}_1^E = V^E - w_2(\frac{dV^E}{dw_2})_{T,P} \tag{6}$$

$$\overline{V}_2^E = V^E - w_1(\frac{dV^E}{dw_1})_{T,P} \tag{7}$$

where $\overline{V}_1^E$ and $\overline{V}_2^E$ are the excess partial volumes (cm^3 g^{-1}) of the water and protein.

3. EXPERIMENTAL

3.1. Materials

Hen egg-white ovalbumin of the highest commercially available purity was purchased from Sigma Chemical Company (St. Louis, MO, US) and was used without further purification. The purity of protein samples was verified through electrophoresis and dynamic light scattering measurements (90Plus Particle Size Analyzer, Brookhaven Instruments Corporation, US) to be more than 95%. The molecular mass of ovalbumin was taken as 45,000 Da. The water was doubly distilled.

3.2. Water Sorption Measurements

The dried protein (zero hydration level) was obtained through vacuum drying using a microthermoanalyser "Setaram" MGDTD-17S (±0.00001 g) at

25 °C and 0.1 Pa, until a constant sample mass was reached. The water content of the dried protein was estimated as 0.003 ± 0.002 g water g^{-1} protein using the Karl Fischer titration method, according to the recommendations [31, 32].

At the lowest and intermediate water mass fractions, the protein preparations were prepared as follows. The initially dehydrated protein samples were exposed to pure water vapour.

The water vapour was consecutively flowed through a thermostated glass tube with drying agent (P_2O_5), a thermostated saturator filled with saturated salt solution, and a cell containing the protein sample. The schematic representation of the experimental setup is given in Reference [24]. An external ethylene glycol thermostat (RC 6 from Lauda, Germany) was used to control the temperature with a precision of 0.1 °C. The sorption equilibrium was reached after 180 min. The water activity (a_w) in the vapour phase was adjusted by changing the saturated salt solution in the saturator. The water activities of the saturated salt solutions were taken from [33, 34].

The following salts were used: LiBr (a_w=0.064), KOH (a_w=0.078), LiCl (a_w=0.11), $CaBr_2$ (a_w=0.17), CH_3COOK (a_w=0.22), $MgCl_2$ (a_w=0.33), K_2CO_3 (a_w=0.44), $Mg(NO_3)_2$ (a_w=0.53), NaCl (a_w=0.75), KCl (a_w=0.84), KNO_3 (a_w=0.94), and K_2SO_4 (a_w=0.97).

The salts used for sample conditioning were of analytical pure grade. The water content of the samples after equilibration was measured by drying under vacuum using a "Setaram" MGDTD-17S microthermoanalyser at 25 °C and 0.1 Pa until a constant sample mass was reached.

3.3. Densitometry

The volumes of the protein-water mixtures, V^m, were calculated using Equation 8:

$$V^{\mathrm{m}} = \frac{1}{d_{\mathrm{m}}} \tag{8}$$

where d_m is the density of the protein-water mixtures, g cm^{-3}.

At the highest water mass fractions (w_1~0.6-1.0), the densities of the protein-water mixtures were measured using a high precision densitometer (DMA-5000 M, Anton Paar, Austria, precision $\pm 1\times10^{-6}$ g cm^{-3}) at 25 °C. The instrument constant was determined by calibrating the densitometer using double-distilled water of a known density. At the lowest and intermediate water mass fractions (w_1 = 0-0.5), the volumes of the protein–water mixtures were determined using a helium pycnometer (AccuPyc 1330, Micromeritics, 1.0 ml cell) at 25 °C. Helium was used as the displacement gas at an equilibration rate of 0.1 kPa min^{-1}. Ten sample runs were sufficient for values of the volumes significant to the third decimal place.

At the lowest and intermediate water mass fractions (w_1 = 0-0.5), the protein preparations were prepared as described in Section 3.2.

4. Results and Discussion

Figure 1 presents the dependence of the volumes of binary ovalbumin-water mixtures, V^m, on the water mass fraction, w_1. At high w_1 values, the volumes of the ovalbumin-water mixtures depend linearly on the water content. However, at low w_1 values, there is a deviation from linearity.

The volumes (densities) of the ovalbumin-water systems were measured in References [13, 20, and 21]. As shown in Figure 1 and Table 1, the V^m and $\overline{V}_2$ values are consistent with the previously published results for ovalbumin [13, 20, 21], indicating that the apparatus and the experimental procedure are adequate to calculate the volumes.

Figure 2 shows the dependence of the excess volumes of the studied binary systems, V^E, on the hydration level of ovalbumin. The V^E values were calculated using Equation 1. In ideal binary mixtures (mixtures of two components, W [water] and P [protein]) the average W-P interactions are the same as the average W-W and P-P interactions. The excess volumes are zero for ideal mixtures. Non-ideal mixtures are composed of particles for which the W-W, P-P, and W-P interactions are all different.

As shown in Figure 2, the excess volumes of binary protein-water mixtures are negative over the whole composition range.

A well-defined minima in the excess volume values exists at w_1~0.1, which indicates the densest structure packing of the studied binary systems. This means that the ovalbumin-water mixtures are non-ideal.

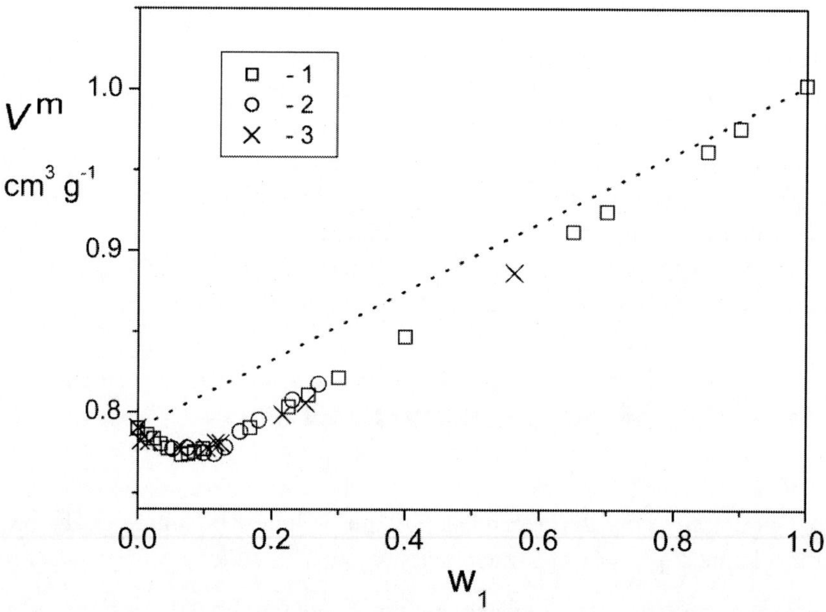

The dashed line corresponds to the ideal binary mixture. (2) Ovalbumin (Adapted data from Reference [20]). (3) Ovalbumin (Adapted data from Reference [21]).

Figure 1. The V^m values for the water-ovalbumin mixtures as a function of the mass fraction of water, w_1, at 25 °C: (1) Ovalbumin (this study). The standard errors of estimation of the V^m values were 0.002-0.004 cm^3 g^{-1}. Each experiment was performed 3-4 times.

Table 1. Partial specific volume of ovalbumin, $\overline{V}_2$ [a)], at w_1=1.0, w_1=0 and 25°C

Protein	$\overline{V}_2$ (w_1=1.0), cm^3 g^{-1}	$\overline{V}_2$ (w_1=0), cm^3 g^{-1}
Ovalbumin	0.746[13]; 0.7477 [20] 0.745 [this study]	0.790 [21]; 0.793 [this study]

[a]The $\overline{V}_2$ values were estimated using the following equation: $V^m = w_1\overline{V}_1 + w_2\overline{V}_2$.

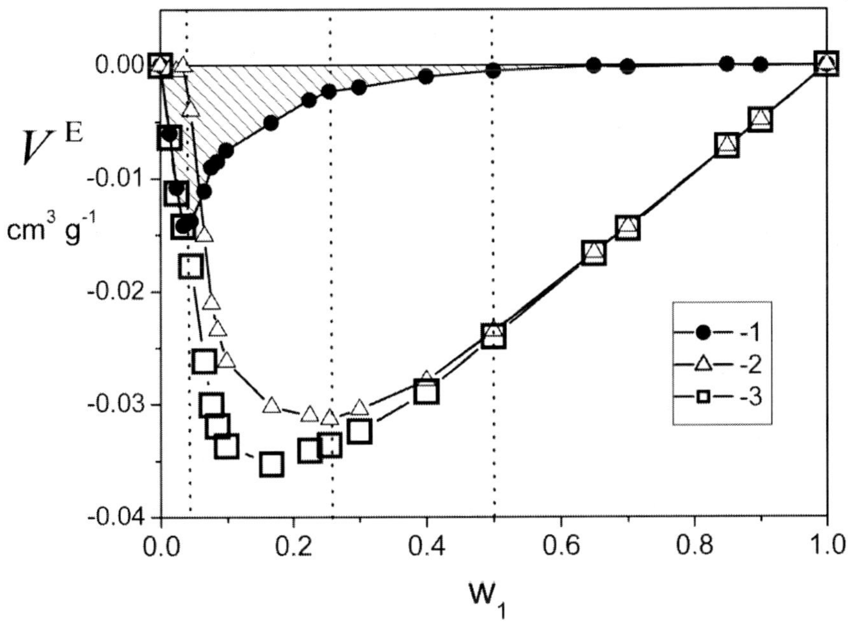

Figure 2. Ovalbumin-water mixtures. Excess thermodynamic functions plotted versus the mass fraction of water, w_1, at 25 °C: (1) $w_1\overline{V}_1^E$; (2) $w_2\overline{V}_2^E$; (3) V^E.

Figure 2 presents the dependencies of the $w_i\overline{V}_i^E$ ($w_1\overline{V}_1^E$ and $w_2\overline{V}_2^E$) functions plotted versus the water mass fraction. These functions were calculated using Equation 3. They show the individual protein and water contributions to the excess thermodynamic functions in the entire range of water content. Figure 2 demonstrates that the protein contribution ($w_2\overline{V}_2^E$) reaches maximal values at w_1~0.25. However, the $w_1\overline{V}_1^E$ function reaches maximal values at low water content (w_1~0.05).

Figure 3 shows the dependence of the $\overline{V}_1^E$ values on the hydration level of ovalbumin. The $\overline{V}_1^E$ values were calculated using Equation 6. Figure 4 presents the $\overline{V}_2^E$ values as a function of the mass fraction of water. The $\overline{V}_2^E$ values were calculated using Equation 7.

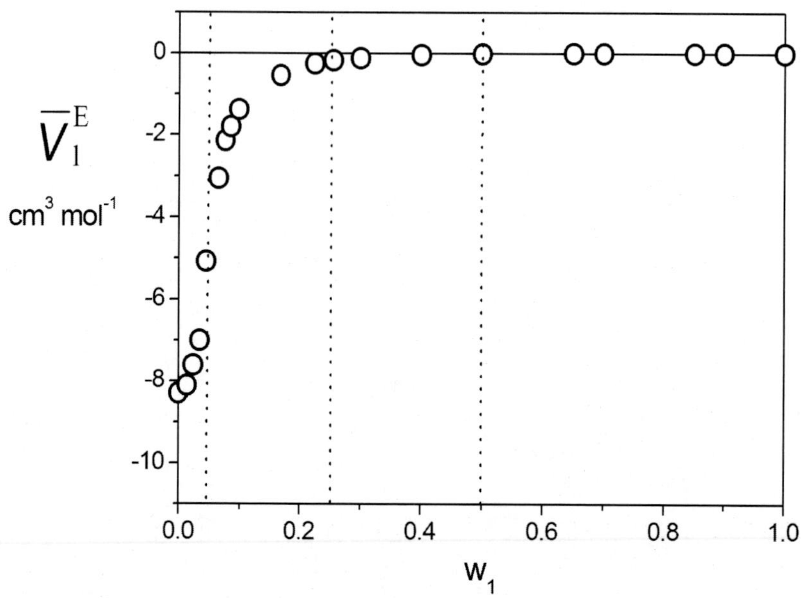

Figure 3. Excess partial volume of water, $\overline{V}_1^E$, as a function of the mass fraction of water, w_1, at 25 °C.

The excess partial volume curves presented in Figures 2-4 can be divided into four parts. At the lowest water content (w_1=0-0.05), the $\overline{V}_2^E$ values are close to zero, reflecting the fact that all of the protein molecules contact the same protein molecules during this range of water content (Figure 4). The proteins are in a glassy (rigid) state [1]. In the glassy state, the dehydration-induced conformational changes and restrictions on conformational transitions cause the protein to become frozen into a broad distribution of conformational states. No biological activity was observed at $w_1<0.05$ (Figure 3 from Reference [7]).

At $w_1<0.05$, the $w_2\overline{V}_2^E$ values are close to zero (Figure 2). At $w_1\sim0.05$, the $w_1\overline{V}_1^E$ values reach maximal values.

These facts imply that, at the lowest water content, changes in the excess volume, V^E, solely reflect water addition.

Hutchens et al. [35] examined the heat capacities of insulin at w_1 = 0 and 0.038 and chymotrypsinogen A at w_1 = 0 and 0.096, from -263 to 37 °C. No phase transition corresponding to the ice-liquid water transition was observed at low hydration.

This region corresponds to the hydration of charged groups [1, 2, 36]. At low water content, the $\overline{V}_1^E$ values differ significantly from zero (Figure 3). The $\overline{V}_1^E$ values are highly negative. As concluded in Reference [8], the $\overline{V}_1^E$ (w_1=0) values are close to those observed for diethylamine, suggesting that, at the lowest w_1 values, the interaction with the protein elements containing the amino groups might be a dominant factor controlling the state of water molecules.

Part 2 (w_1=0.05-0.25) corresponds to the hydration of polar groups [1, 2, 36]. In this water content range, the $\overline{V}_1^E$ values change sharply from highly negative to moderate values (Figure 3). This sharp transition reflects the formation of a spanning hydrogen-bonded network of water at the protein surface [2, 4]. The formation of this network occurs via a quasi-two-dimensional percolation transition of the hydration water at the enzyme surface [4].

The results obtained for biopolymers using several experimental methods have been summarised in Reference [1]. The results suggest that proteins undergo a glasslike dynamic transition at a water content of approximately 10 wt% at 25 °C. This water content falls within part 2 in the present study.

The transition from the glassy (rigid) to the flexible (elastic) state is accompanied by significant changes in the properties of the proteins [1].

For example, the apparent heat capacity of lysozyme, ϕC_{p2}, determined from isothermal experiments using a drop calorimeter, increases from extremely low values to high values in this water content interval [37]. During the isothermal sorption of water, a glasslike transition results in a step on the excess partial volume of ovalbumin (Figure 4). The $\overline{V}_2^E$ values change sharply from extremely low values to highly negative values.

Part 3 (w_1 = *0.25-0.5)* indicates the appearance of the "structured" water [36]. "Structured" water is composed of molecules that interact with hydrophobic surface patches on the protein-bridging between bound water molecules.

Rupley and Careri [37] asserted that this region contains condensed water molecules over weakly interacting surface elements, which are likely nonpolar atoms, not polar groups or adjacent-charged.

The onset of biological activity was observed in this region. The enzyme-catalysed hydrolysis of the hexasaccharide of N-acetylglucosamine [(GlcNAc)$_6$] ([2], Figure 4 from Reference [8]) was examined as a function of the water content and showed that the reaction grows sharply at $w_1 > 0.15$.

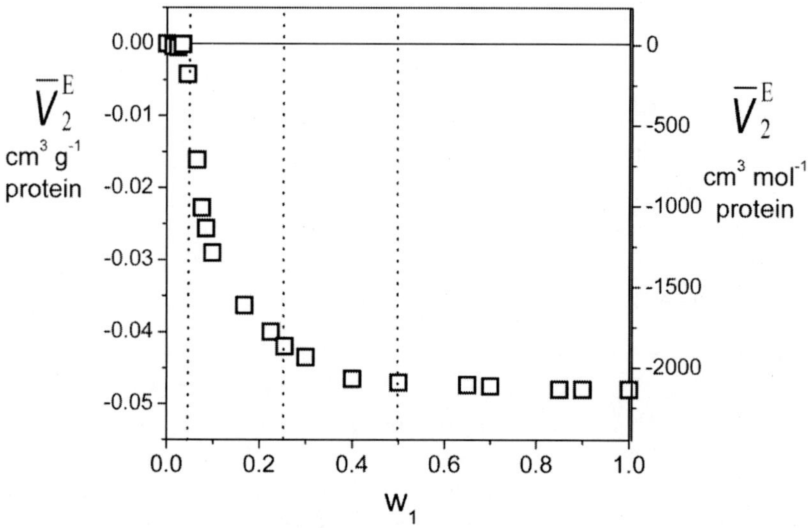

Figure 4. Excess partial volume of ovalbumin, $\overline{V}_2^{\mathrm{E}}$, as a function of the mass fraction of water, w_1, at 25 °C.

The biological activity was suppressed within parts 1 and 2 (Figure 4 from Reference [8]).

At the highest water content ($w_1 > 0.5$), the proteins are in a flexible (elastic) state [1]. Excess partial quantities attain their fully hydrated values. The $\overline{V}_2^E$ values (Figure 4) reach minimal values. At $w_1<0.5$, the $w_1\overline{V}_1^E$ values are near zero (Figure 2). Bulk water was observed in this region through proton NMR measurements [36].

The molecular motion of these water molecules is solely determined through the interaction characteristics of water molecules. Water addition has no significant effect on the excess thermodynamic functions. At $w_1 > 0.5$, the changes in the excess thermodynamic functions primarily reflect changes in the state of ovalbumin.

To show the predictive ability of the proposed methodology, we related the obtained results to the general properties of the proteins. Hydrophobicity and polarity were selected as possible common parameters. The hydrophobicity of the studied proteins has been estimated as the average hydrophobicity of the side chains of the constitutive amino acids, H_ϕ (J per residue). The calculated H_ϕ values for many proteins were collected in [38].

We calculated the areas under the $w_1\overline{V}_1^E = f(w_1)$ and $w_2\overline{V}_2^E = f(w_1)$ curves for ovalbumin [this study], lysozyme, β-lactoglobulin, ribonuclease A, and serum albumin [8, 9]. These areas were defined as $A(w_1\overline{V}_1^E)$ and $A(w_2\overline{V}_2^E)$, respectively. The $A(w_1\overline{V}_1^E)$ area for ovalbumin is shown shaded in Figure 2.

The $A(w_2\overline{V}_2^E)$ values are plotted as a function of H_ϕ in Figure 5. There is a clear correlation between the $A(w_2\overline{V}_2^E)$ values and H_ϕ. This correlation shows that the more hydrophobic a protein is, the more significant the $A(w_2\overline{V}_2^E)$ value is.

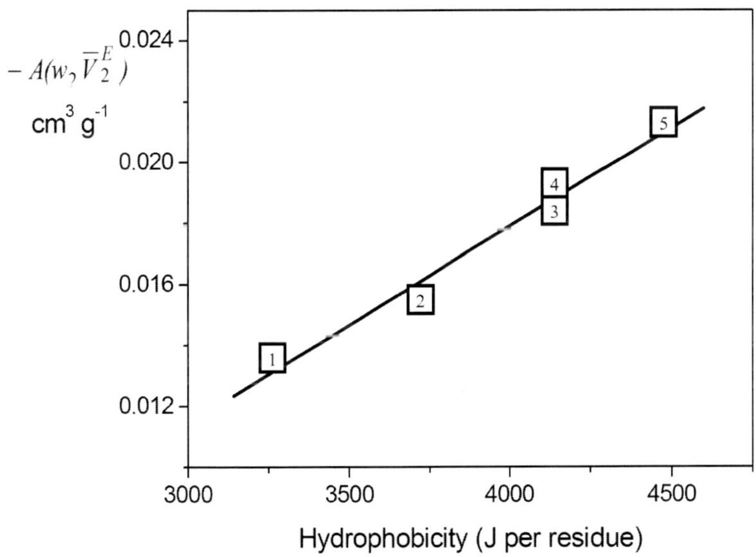

Relationship between the $A(w_2\overline{V}_2^E)$ value and the hydrophobicity, H_ϕ, of proteins.

$-A(w_2\overline{V}_2^E) = 0.008(0.002) + 6.5\cdot10^{-6}(0.6\cdot10^{-6})*H_\phi$, where the number of the experimental points is N=5, the standard error of estimation is S_o=6.0*10^{-4}, and the correlation coefficient is R=0.985.

Figure 5. (1) RNase A; (2) Lysozyme; (3); Ovalbimin; (4) Serum albumin; (5) β-Lactoglobulin.

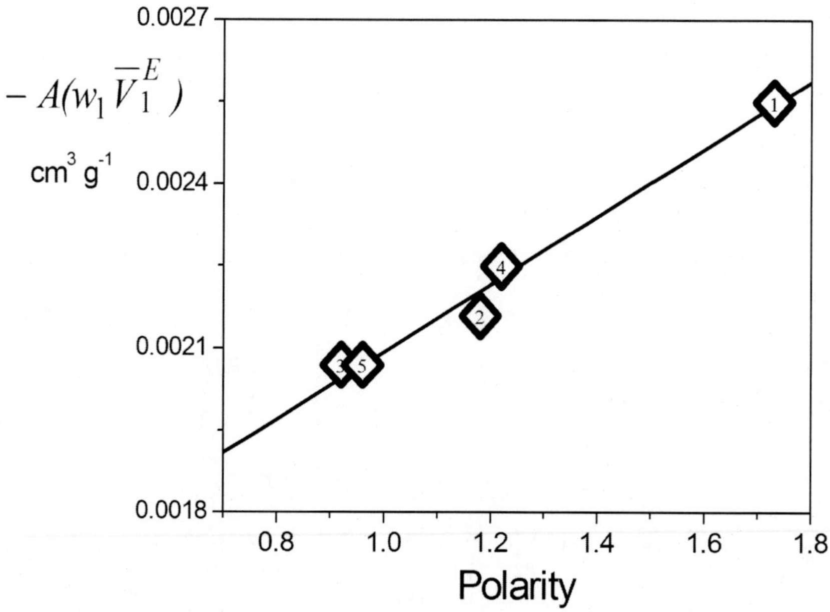

Relationship between the $A(w_1\overline{V}_1^E)$ value and Bigelow's polarity parameter, *P*, of proteins.

$-A(w_1\overline{V}_1^E) = 1.5\cdot10^{-3}(0.6\cdot10^{-5}) + 6.1\cdot10^{-4}(0.5\cdot10^{-4})*P$, where the number of the experimental points is N=5, the standard error of estimation is S_o=3.2*10^{-5}, and the correlation coefficient is R=0.99.

Figure 6. (1) RNase A; (2) Lysozyme; (3) Ovalbimin; (4) Serum albumin; (5) β-Lactoglobulin.

A polarity parameter, *P*, was proposed by Bigelow [39]. It was defined as a volume ratio of polar amino acids to non-polar amino acids. There is a good correlation between the $A(w_1\overline{V}_1^E)$ values and Bigelow's polarity parameter (Figure 6). This correlation means that the more hydrophilic a protein is, the more significant the $A(w_1\overline{V}_1^E)$ value is.

As a rule, most hydrophobic groups of globular proteins are buried in the interior of the protein molecule. On the other hand, charged and polar residues are located preferentially on the surface. H_ϕ may be regarded as a parameter characteristic mostly of the hydrophobic interior of a protein. *P* reflects the degree of the polarity of the protein surface.

RNase A is the most hydrophilic protein in this study. As shown in Figure 6, this protein has the most significant hydrophilic hydration contribution, $A(w_1\overline{V}_1^E)$. Lactoglobulin is the most hydrophobic protein. As shown in Figure 5, this protein has the most significant hydrophobic hydration contribution, $A(w_2\overline{V}_2^E)$. These correlations show that the contributions corresponding to the hydration of the protein interior and the protein surface are coupled differently to the excess volumes. The changes in the partial quantities corresponding to the protein interior are largely described by the $A(w_2\overline{V}_2^E)$ values. The changes in the partial quantities corresponding to the protein surface are described by the $A(w_1\overline{V}_1^E)$ values.

References

[1] Gregory, R. B. (1995) Protein hydration and glass transition behavior. In: Gregory, R. B., editor. *Protein-Solvent Interactions*, Marcel Dekker, New York, 191-264.

[2] Rupley, J. A., Careri, G. (1991) Protein hydration and function. *Adv. Protein Chem.*, 41, 37-172.

[3] Kuntz, I. D., Kauzmann, W. (1974) Hydration of proteins and polypeptides. *Adv. Protein. Chem.*, 28, 239-345.

[4] Oleinikova, A., Smolin, N., Brovchenko, I., Geiger, A., Winter, R. (2005) Properties of spanning water networks at protein surfaces. *J. Phys. Chem. B*, 109, 1988-1998.

[5] Sirotkin, V. A., Solomonov, B. N., Faizullin, D. A., Fedotov, V. D. (2002) Sorption of water and acetonitrile by human serum albumin. *Russ. J. Phys. Chem. A.* 76: 2051-2057.

[6] Durchschlag, H., Zipper, P. (2001) Comparative investigations of biopolymer hydration by physicochemical and modeling techniques. *Biophys. Chem.*, 93, 141-157.

[7] Sirotkin, V. A., Khadiullina, A. V. (2011) Hydration of proteins: excess partial enthalpies of water and proteins. *J. Phys. Chem. B*, 115, 15110-15118.

[8] Sirotkin, V. A., Komissarov, I. A., Khadiullina, A. V. (2012) Hydration of proteins: excess partial volumes of water and proteins, *J. Phys. Chem. B*, 116, 4098-4105.

[9] Sirotkin, V. A., Khadiullina, A. V. (2014) A study of the hydration of ribonuclease A using densitometry: Effect of the protein hydrophobicity and polarity. *Chem. Phys. Lett.*, 603, 13-17.

[10] Chalikian, T. V., Totrov, M., Abagyan, R., Breslauer, K. J. (1996) The hydration of proteins as derived from volume and compressibility measurements: Cross correlating thermodynamic and structural data. *J. Mol. Biol.*, 260, 588-603.

[11] H. Seemann, H., Winter, R., Royer, C. (2001) Volume, expansivity and isothermal compressibility changes associated with temperature and pressure unfolding of staphylococcal nuclease. *J. Mol. Biol.*, 307, 1091-1102.

[12] Sirotkin, V. A., Winter, R. (2010) Volume changes associated with guanidine hydrochloride, temperature, and ethanol induced unfolding of lysozyme. *J. Phys. Chem. B*, 114, 16881-16886.

[13] Gekko, K., Hasegawa, Y. (1986) Compressibility-structure relationship of globular proteins. *Biochemistry*, 25, 6563-6571.

[14] Svergun, D., Richard, S., Koch, M. H., Sayers, Z., Kuprin, S., Zaccai, G. (1998) Protein hydration in solution: Experimental observation by x-ray and neutron scattering. *Proc. Natl. Acad. Sci. US* 25, 2267-2272.

[15] Gerstein, M., Chothia, C. (1993) Packing at the protein-water interface. *Proc. Natl. Acad. Sci. US* 93, 10167-10172.

[16] Merzel, F., Smith, J. C. (2002) Is the first hydration layer of lysozyme of higher density than bulk water? *Proc. Natl. Acad. Sci. US* 99, 5378-5383.

[17] Merzel, F., Smith, J. C. (2005) High-density hydration layer of lysozymes: Molecular dynamics decomposition of sorption scattering data. *J. Chem. Inf. Model*, 45, 1593-1599.

[18] Kim, K., Kauzmann, W. (1980) Concentration dependence of the specific volume of human oxyhemoglobin, bovine serum albumin, ovalbumin solutions. *J. Phys. Chem.*, 84, 163-165.

[19] Bernhardt, J., Pauly, H. (1980) Pycnometric vs. densimetric measurements of highly viscous protein solutions. *J. Phys. Chem.*, 84, 158-162.

[20] Bull, B. H., Breese, K. (1968) Protein hydration. II. Specific heat of egg albumin. *Arch. Biochem. Biophys.*, 128, 497-502.

[21] Neurath, H., Bull, H. B. (1936) Denaturation and hydration of proteins. *J. Biol. Chem.*, 115, 519-528.

[22] Low, B. W., Richards, F. M. J. (1954) Measurements of the density, composition and related unit cell dimensions of some protein crystals. *J. Amer. Chem. Soc.*, 76, 2511-2518.

[23] McMeekin, T. L., Groves, M. L., Hipp, N. J. (1954) Partial specific volume of the protein and water in beta-lactoglobulin crystals, *J. Polymer Sci.*, 12, 309-315.

[24] Sirotkin, V. A., Khadiullina, A. V. (2013) Gibbs energies, enthalpies, and entropies of water and lysozyme at the inner edge of excess hydration. *J. Chem. Phys.*, 139, 075102/1-075102/9.

[25] Yang, Y., Barendregt, A., Kamerling, J. P., Heck, A. J. R. (2013) Analyzing protein micro-heterogeneity in chicken ovalbumin by high-resolution native mass spectrometry exposes qualitatively and semi-quantitatively 59 proteoforms. *Anal. Chem.*, 85, 12037-12045.

[26] Nisbet, A. D., Saundry, R. H., Moir, A. J. G., Fothergill, L. A., Fothergill, J. E. (1981) The complete amino-acid sequence of hen ovalbumin. *Eur. J. Biochem.*, 115, 335-345.

[27] Huntington, J. A., Stein, P. E. (2001) Structure and properties of ovalbumin. *J. Chromatog. B*, 756, 189-198.

[28] Atkins, P. W. (2006) *Physical Chemistry*. 8th ed. Oxford: Oxford University Press.

[29] Prausnitz, J. M. (1969) *Molecular Thermodynamics of Fluid-Phase Equilibria*. N.J.: Prentice-Hall, Inc., Engelwood Cliffs.

[30] Belousov, V. P., Panov, M. Y. (1994) *Thermodynamic properties of aqueous solutions of organic substances*. Boca Raton, Fla.: CRC Press.

[31] Borisover, M. D., Sirotkin, V. A., Solomonov, B. N. (1995) Thermodynamics of water binding by human serum albumin suspended in acetonitrile. *Thermochim. Acta*, 254, 47-53.

[32] Sirotkin, V. A., Zinatullin, A. N. Solomonov, B. N., Faizullin, D. A., Fedotov, V. D. (2002) Interaction enthalpies of solid bovine pancreatic chymotrypsin with organic solvents: comparison with FTIR-spectroscopic data. *Thermochim. Acta*, 382, 151-160.

[33 Greenspan, L. (1977) Humidity fixed points of binary saturated aqueous solutions. *J. Res. Natl. Bur. Stand.*, Sect. A 81, 89.

[34] Nikolskii, B. P. (1963) *Spravochnik Khimika (Chemist's Handbook)* Leningrad, Russia: Goskhimizdat.

[35] Hutchens, J. O., Cole, A. G., Stout, J. W. (1969) Heat capacities from 11 to 305 K and entropies of hydrated and anhydrous bovine zinc insulin and bovine chymotrypsinogen A. *J. Biol. Chem.*, 244, 26-32.

[36] Fullerton, G. D., Ord, V. A., Cameron, I. L. (1986) An evaluation of the hydration of lysozyme by an NMR titration method. *Biochem. Biophys. Acta*, 869, 230-246.
[37] Yang, P.-H., Rupley, J. A. (1979) Protein-water interactions. Heat capacity of the lysozyme-water system. *Biochemistry*, 18, 2654-2661.
[38] Bigelow, C. C., Channon, M. (1976) *Handbook of Biochemistry and Molecular Biology, Proteins.* Cleveland, OH: CRC Press.
[39] Bigelow, C. C. (1967) On the average hydrophobicity of proteins and the relation between it and protein structure. *J. Theor. Biol.,* 16, 187-211.

In: Protein – Water Interactions
Editor: Vladimir A. Sirotkin ISBN: 978-1-63463-007-8
© 2014 Nova Science Publishers, Inc.

Chapter 4

ANALYSIS OF HYDRATION OF OVALBUMIN BY ISOTHERMAL CALORIMETRY

Vladimir A. Sirotkin*, and Aigul V. Khadiullina
Kazan Federal University, A.M. Butlerov Institute of Chemistry, Kazan, Russia

ABSTRACT

Isothermal calorimetry was applied to characterize the hydration dependencies of the excess thermodynamic functions of binary protein-water systems. Here, in particular, the excess partial enthalpies of water and hen egg-white ovalbumin have been determined. The excess partial enthalpies for ovalbumin are compared with the published data for several unrelated globular proteins (ribonuclease A, lysozyme, chymotrypsinogen A, serum albumin, lactoglobulin). These biomacromolecules represent a series of proteins in which the hydrophobicity of proteins is gradually changed over a wide range. It was found that the excess partial quantities for the studied proteins are determined by the hydration of the hydrophilic and hydrophobic protein groups. The more hydrophilic a protein, the more significant a hydrophilic hydration contribution is and *vice versa*. Ribonuclease A is the most hydrophilic protein in the present study. This protein has the most significant hydrophilic hydration contribution. Lactoglobulin is the

* Email address: vsir@mail.ru (V.A. Sirotkin).

most hydrophobic protein under the study. This protein has the most significant hydrophobic hydration contribution. Ovalbumin shows intermediate properties.

Keywords: Protein (biomacromolecule) hydration, Biocalorimetry, Excess functions, Enthalpy

1. INTRODUCTION

Thermochemical studies have traditionally been of great importance in ascertaining a better understanding of protein-water interactions [1-7]. Herein, we present a review of the available literature concerning the hydration of proteins. Because our current research involves a thermochemical study of protein-water systems, this literature review will focus on studies concerning the corresponding enthalpy changes. More comprehensive reviews have been previously published in [1-4].

Isothermal calorimetry is one of the most effective methods for obtaining reliable thermochemical information on the interactions of proteins with water in various environments. Amberg measured the heats of water vapor sorption using bovine serum albumin within a water content range of 0-12 % (g g^{-1}) at 20 °C [8]. The enthalpies of solution of dried and hydrated ribonuclease A (RNase A) was measured using isothermal calorimetry at 25 °C [9]. The water content of RNase A varied from 0 to 26 % (g g^{-1}), and the enthalpy of the solution was negative (exothermic). As the water content of RNase A increases, the enthalpy of the solution approaches zero. Smith et al. calorimetrically measured the heat of water adsorption by lysozyme in the range of relative water vapor pressures from 0 to 0.895 [10] and obtained both the sorption isotherm and enthalpy of hydration of the enzyme within a water content range of 0-18 % (g g^{-1}) at 25 °C. Sorption calorimetry has been used to measure the adsorption isotherm of water by lysozyme and the corresponding heat effects in the entire range of water activity [11]. Our research group has developed an experimental method for measuring the heat effects of the protein hydration [12]. The interaction enthalpy depends significantly on the initial water content and hydration history. The most important observations of these calorimetric studies [8-12] can summarized as follows: The hydration enthalpies vary strongly with the hydration level. They are large at low hydration levels. As the water content increases, the hydration enthalpies approach the enthalpy of condensation of pure water.

The measured enthalpies [8-12] contain total information on the binary water-protein systems, including the corresponding conformational changes in the protein structure and glass transition.

However, the simultaneous estimation of the protein and water contributions to the enthalpy (or other thermodynamic functions) of binary protein-water systems in the entire range of water content has not been attempted [1-4, 8-12].

In our previous studies [13-16], we proposed a novel method to investigate the protein-water interactions. This method is based on the analysis of the thermodynamic functions of mixing. It facilitates the individual evaluation of the protein and water partial quantities in the entire range of water content. The present study is aimed to understand what molecular parameters control the excess enthalpies of the binary protein-water systems.

To test the predictive ability of the proposed method, a system containing, as a model protein, hen egg-white ovalbumin was studied in this work. The excess partial quantities for ovalbumin were compared with the published data for several unrelated globular proteins (lysozyme, chymotrypsinogen A, serum albumin, lactoglobulin) [13,16]. These biomacromolecules represent a series of proteins in which the hydrophobicity of proteins is gradually changed over a wide range.

Ovalbumin is the main protein found in egg white [17-19]. This protein consists of 385 amino acids residues. It is generally assumed to be a storage protein.

2. Methodology

2.1. Excess Partial Quantities

The thermodynamic properties of a real binary water-protein system can be expressed in terms of the excess functions, Z^{E} [20-22], i.e., the difference between the observed thermodynamic function of mixing, Z^{m}, and the function for an ideal binary mixture, $Z^{\mathrm{m}}_{\mathrm{id}}$:

$$Z^E = Z^m - Z^m_{id} \tag{1}$$

The Z_{id}^{m} values can be calculated using Eq. 2:

$$Z_{id}^{m} = w_1 Z_1^0 + w_2 Z_2^0 \tag{2}$$

where Z_1^0 and Z_2^0 are the thermodynamic function values for pure water and pure protein, and w_1 and w_2 are the mass fractions of water and protein, respectively.

Deviations of the excess functions from zero indicate the extent to which the studied binary system is non-ideal due to strong specific interactions between components (i.e., hydrogen bonding and charge-charge interactions).

The Z^E values are composed of two components:

$$Z^E = w_1 \overline{Z}_1^E + w_2 \overline{Z}_2^E \tag{3}$$

where $\overline{Z}_1^E$ is the excess partial function for component 1 (water), and $\overline{Z}_2^E$ is the excess partial function for component 2 (protein).

2.2. Excess Partial Enthalpies

The excess partial enthalpies of the protein and water were calculated as described previously [13,16]. This method facilitates the estimation of the individual protein and water partial quantities in the entire range of water content. The interaction enthalpy observed during the addition of water to the protein is considered to be an excess quantity of the mixing of water with the protein (H^{E}, Eq. 3). The H^{E} value is calculated from the enthalpies of the solution of the dried and hydrated protein in pure liquid water. The enthalpy of the solution of the dried protein in pure liquid water, $\Delta_{sol}\mathrm{H}(dried)$, is described in Eq. 4. The final state of the protein (Eqs. 4 and 5) is a diluted solution in water at 25°C and at atmospheric pressure:

$$\Delta_{sol}\mathrm{H}(dried) = [\overline{\mathrm{H}}_{P}^{sol}\mathrm{m}_{P}] - [\overline{\mathrm{H}}_{P}^{o}\mathrm{m}_{P}] + [\overline{\mathrm{H}}_{w,.protein}^{sol}\mathrm{m}_{w}^{sol}] - [\overline{\mathrm{H}}_{w,.liquid}\mathrm{m}_{w}^{sol}] \quad (4)$$

where $\overline{H}_{\mathrm{P}}^{0}$ is the partial enthalpy of the dried protein at 25°C and at atmospheric pressure; $\overline{H}_{\mathrm{w,.liquid}}$ is the partial enthalpy of water in pure liquid state at 25°C and at atmospheric pressure; $\overline{H}_{\mathrm{w,.protein}}^{\mathrm{sol}}$ is the partial enthalpy of water bound to the protein in the diluted aqueous solution; $\overline{H}_{\mathrm{P}}^{\mathrm{sol}}$ is the partial enthalpy of the protein in the solution; m_{P} is the amount of the protein; $m_{\mathrm{W}}^{\mathrm{h}}$ is the amount of water bound to the protein; and $m_{\mathrm{W}}^{\mathrm{sol}}$ is the amount of water bound to the protein in the diluted aqueous solution.

The enthalpy of solution of the initially-hydrated protein in pure liquid water, $\Delta_{sol}\mathrm{H}(hydrated)$, is described in Eq. 5:

$$\Delta_{sol}\mathrm{H}(hydrated) = [\overline{\mathrm{H}}_{P}^{sol}\mathrm{m}_{P}] - [\overline{\mathrm{H}}_{P}^{h}\mathrm{m}_{P}] + [\overline{\mathrm{H}}_{w,.protein}^{sol}\mathrm{m}_{w}^{sol}] - [\overline{\mathrm{H}}_{w,.liquid}\mathrm{m}_{w}^{sol}] + [\overline{\mathrm{H}}_{w,.liquid}\mathrm{m}_{w}^{h}] - [\overline{\mathrm{H}}_{w,.protein}^{h}\mathrm{m}_{w}^{h}] \quad (5)$$

where $\overline{H}_{\mathrm{w,.protein}}^{\mathrm{h}}$ is the partial enthalpy of water bound to the protein, and $\overline{H}_{\mathrm{P}}^{\mathrm{h}}$ is the partial enthalpy of the hydrated protein.

As shown in Eq. 6, the H^{E} value is calculated from the enthalpies of solution of the dried, $\Delta_{sol}\mathrm{H}(dried)$, and hydrated, $\Delta_{sol}\mathrm{H}(hydrated)$, protein in water, measured using isothermal calorimetry:

$$\frac{\Delta_{sol}\mathrm{H}(dried) - \Delta_{sol}\mathrm{H}(hydrated)}{\mathrm{m}_w^h + \mathrm{m}_P} =$$

$$= [\overline{\mathrm{H}_P^h} - \overline{\mathrm{H}_P^0}]\frac{\mathrm{m}_P}{\mathrm{m}_P + \mathrm{m}_w^h} +$$

$$[\overline{\mathrm{H}}_{w,.protein}^h - \overline{\mathrm{H}}_{w,.liquid}]\frac{\mathrm{m}_w^h}{\mathrm{m}_P + \mathrm{m}_w^h} = \tag{6}$$

$$= [\overline{\mathrm{H}_P^h} - \overline{\mathrm{H}_P^0}]\mathrm{w}_2 + [\overline{\mathrm{H}}_{w,.protein}^h - \overline{\mathrm{H}}_{w,.liquid}]\mathrm{w}_1 = \mathrm{H}^E$$

The excess partial enthalpies of the protein and water are calculated using Eqs. 7 and 8:

$$\overline{\mathrm{Z}}_1^E = \mathrm{Z}^E - \mathrm{w}_2(\frac{d\mathrm{Z}^E}{d\mathrm{w}_2})_{T,P} =$$

$$[\overline{\mathrm{H}}_{w,.protein} - \overline{\mathrm{H}}_{w,.liquid}] = \overline{\mathrm{H}}_1^E \tag{7}$$

$$\overline{Z}_2^{\mathrm{E}} = Z^{\mathrm{E}} - w_1(\frac{dZ^{\mathrm{E}}}{dw_1})_{\mathrm{T,P}} = [\overline{H_{\mathrm{P}}^{\mathrm{h}}} - \overline{H_{\mathrm{P}}^{0}}] = \overline{H}_2^{\mathrm{E}} \tag{8}$$

where $\overline{H}_1^{\mathrm{E}}$ (J g^{-1} water) and $\overline{H}_2^{\mathrm{E}}$ (J g^{-1} protein) are the excess partial enthalpies of the water and protein, respectively, and w_1 and w_2 are the mass fractions of water and protein, respectively. The symbols used in section "Methodology" are summarized in Table 1.

Table 1. Symbols Used in Section "Methodology"

Symbol	Dimension	Name
W	-	Water
P	-	Protein
h	-	Hydrated
$m_{\mathrm{W}}^{\mathrm{h}}$	g water	Amount of water bound to the protein
m_{P}	g protein	Amount of the protein
$m_{\mathrm{W}}^{\mathrm{sol}}$	g water	Amount of water bound to the protein in the diluted aqueous solution
W_1	-	Mass fraction of water
W_2	-	Mass fraction of protein
$\overline{H}_{\mathrm{w,.liquid}}$	J g^{-1} water	Partial enthalpy of water in pure liquid state at 25°C and atmospheric pressure
$\overline{H}_{\mathrm{w,.protein}}^{\mathrm{h}}$	J g^{-1} water	Partial enthalpy of water bound to the protein at 25°C and atmospheric pressure
$\overline{H}_{\mathrm{P}}^{0}$	J g^{-1} protein	Partial enthalpy of the dried protein at 25°C and atmospheric pressure
$\overline{H}_{\mathrm{P}}^{\mathrm{h}}$	J g^{-1} protein	Partial enthalpy of the hydrated protein at 25°C and atmospheric pressure
$\overline{H}_{\mathrm{w,..protein}}^{\mathrm{sol}}$	J g^{-1} water	Partial enthalpy of water bound to the protein in the diluted aqueous solution
$\overline{H}_{\mathrm{P}}^{\mathrm{sol}}$	J g^{-1} protein	Partial enthalpy of the protein in the diluted aqueous solution
$\Delta_{sol} H\,(dried)$	J	Enthalpy of solution of the dried protein in pure water at 25°C and atmospheric pressure
$\Delta_{sol} H\,(hydrated)$	J	Enthalpy of solution of the hydrated protein in pure water at 25°C and atmospheric pressure

3. EXPERIMENTAL

3.1. Materials

Hen egg-white ovalbumin of the highest commercially available purity was purchased from Sigma Chemical Company (St. Louis, MO, USA) and was used without further purification. The purity of protein samples was verified through electrophoresis to be more than 95%. The molecular mass of ovalbumin was taken as 45,000 Da. The water was doubly distilled.

3.2. Water Sorption Measurements

The dried protein preparation (zero hydration level) was obtained through vacuum drying using a microthermoanalyzer "Setaram" MGDTD-17S (±0.00001 g) at 25 °C and 0.1 Pa, until a constant sample mass was reached. The water content of the dried protein was estimated as 0.003 ± 0.002 g water g^{-1} protein using the Karl Fischer titration method, according to the recommendations [23]. This zero hydration level was described using $\overline{H}_P^0$ (Eqs. 4-6).

At the lowest and intermediate water mass fractions, the protein preparations were prepared as follows. The initially dehydrated protein samples were exposed to pure water vapor. The water vapor was consecutively flowed through a thermostated glass tube with drying agent (P_2O_5), a thermostated saturator filled with saturated salt solution, and a cell containing the protein sample. The schematic representation of the experimental setup is given in Figure 1 from Ref [16]. An external ethylene glycol thermostat (RC 6 from Lauda, Germany) was used to control the temperature with a precision of 0.1 °C. Sorption equilibrium was reached after 180 min. The water activity (a_w) in the vapor phase was adjusted by changing the saturated salt solution in the saturator. The water activities of the saturated salt and KOH solutions were taken from [24,25]. The following substances were used: LiBr (a_w=0.064), KOH (a_w=0.078), LiCl (a_w=0.11), $CaBr_2$ (a_w=0.17), CH_3COOK (a_w=0.22), $MgCl_2$ (a_w=0.33), K_2CO_3 (a_w=0.44), $Mg(NO_3)_2$ (a_w=0.53), NaCl (a_w=0.75), KCl (a_w=0.84), KNO_3 (a_w=0.94), and K_2SO_4 (a_w=0.97). The salts used for sample conditioning were of analytical pure grade. The water content of the samples after equilibration was measured by drying under vacuum using a microthermoanalyzer "Setaram" MGDTD-17S at 25 °C and 0.1 Pa, until a constant sample mass was reached.

3.3. Calorimetry

The enthalpies of the solution of the dried, $\Delta_{sol}H(dried)$, and hydrated, $\Delta_{sol}H(hydrated)$, protein in pure liquid water (Eqs. 4 and 5) were measured at 25 °C using a Setaram BT-2.15 calorimeter, according to a previously described procedure [26,27]. Briefly, 8-10 mg of protein was prepared in 4.0 ml of water in the calorimetric cell. A typical duration of a calorimetric experiment lasted approximately 40 min. The typical heat evolution curve recorded from the dissolution of a solid protein in pure liquid water has been previously described [27].

The calorimeter was calibrated using the Joule effect and tested by measuring the solution enthalpy of potassium chloride in water. Potassium chloride of 99.9% purity was recrystallized and dried as previously described [28]. The measured value of the solution enthalpy, $\Delta_{sol}H^{KCl/H_2O}$ (25 °C, m = 0.0277 mol kg^{-1}) = 17.43 ± 0.08 kJ mol^{-1}, is consistent with the recommendations of the ICTAC working group "thermochemistry" [29]: $\Delta_{sol}H^{KCl/H_2O}$ (25 °C, m = 0.02775 mol kg^{-1}) = 17.47 ± 0.07 kJ mol^{-1}.

The dried protein preparation (zero hydration level) was obtained through vacuum drying using a microthermoanalyzer "Setaram" MGDTD-17S at 25 °C and 0.1 Pa, until a constant sample mass was reached. This zero hydration level was described using the $\overline{H}^{0}_{P}$ values (Eqs. 4-6).

At the lowest and intermediate water content, the protein preparations were prepared as described in subsection "Water sorption measurements". At the highest water content, the protein and water samples were mixed in the calorimetric cell at various water mass fractions and at 25 °C. The masses of the protein samples ranged from 8-10 mg. The water content of the samples after equilibration was measured through vacuum drying using a microthermoanalyzer "Setaram" MGDTD-17S at 25 °C and 0.1 Pa, until a constant sample mass was reached.

The hydrated protein samples were described using $\overline{H}^{h}_{P}$ and $\overline{H}_{w,.protein}$ (Eqs. 5-8).

4. Results and Discussion

4.1. Partial Enthalpies of Water and Ovalbumin

Figure 1 shows the dependency of the H^{E} values on the hydration level of ovalbumin. The H^{E} values were estimated using Eq. 6. Figure 2 shows the dependency of the $\overline{H}_1^{\mathrm{E}}$ values on the hydration level of ovalbumin. The $\overline{H}_1^{\mathrm{E}}$ values were calculated using Eq. 7. Figure 3 presents the $\overline{H}_2^{\mathrm{E}}$ values as a function of the mass fraction of water. The $\overline{H}_2^{\mathrm{E}}$ values were calculated using Eq. 8.

In ideal binary mixtures (mixtures of two components, W [water] and P [protein]) the average W-P interactions are the same as the average W-W and P-P interactions. Non-ideal mixtures are composed of particles for which the W-W, P-P, and W-P interactions are all different. As shown in Figure 1, the H^{E} values differ significantly from zero, indicating that the water-protein mixtures are non-ideal in the entire range of water content. Figure 1 present the dependencies of the $w_i\overline{Z}_i^E$ ($w_1\overline{H}_1^{\mathrm{E}}$ and $w_2\overline{H}_2^{\mathrm{E}}$) functions plotted versus the water mass fraction. These functions were calculated using Eq. 3. They show the individual protein and water contributions to the excess thermodynamic functions in the entire range of water content. Figure 1 demonstrates that the protein contribution ($w_2\overline{H}_2^E$) reaches maximal values at w_1~0.25. However, the $w_1\overline{H}_1^E$ function reaches maximal values at low water content (w_1~0.05).

The excess partial enthalpy curves presented in Figs. 1-3 can be divided into four parts.

At the lowest water content *(w_1=0-0.05)*, the $\overline{H}_2^{\mathrm{E}}$ values are close to zero and do not depend on the water content (Figure 3). The proteins are in a glassy (rigid) state [1].

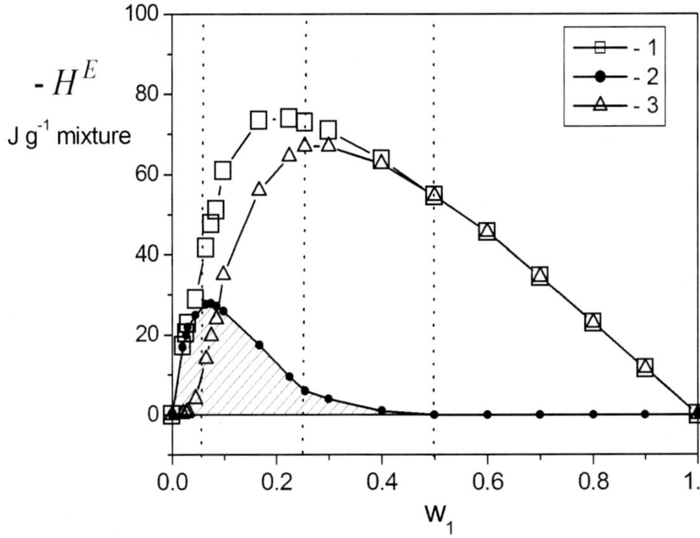

Figure 1. Ovalbumin-water mixtures. Excess thermodynamic functions plotted versus the mass fraction of water, w_1, at 25 °C: (1) H^E; (2) $w_1\overline{H}_1^E$; (3) $w_2\overline{H}_2^E$.

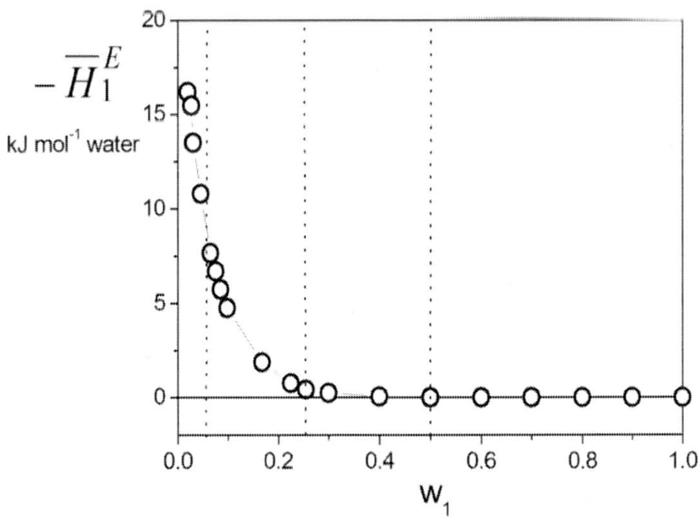

Figure 2. Excess partial enthalpy of water, $\overline{H}_1^E$, as a function of the mass fraction of water, w_1, at 25 °C. The reference state is pure liquid water at 25 °C and atmospheric pressure.

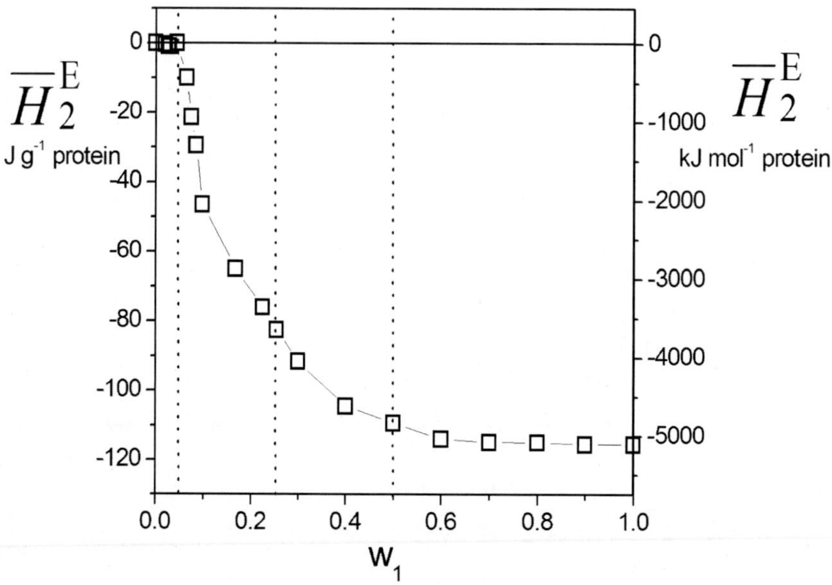

Figure 3. Excess partial enthalpy of ovalbumin, $\overline{H}_2^E$, as a function of the mass fraction of water, w_1, at 25 °C.

In the glassy state, the dehydration-induced conformational changes and restrictions on conformational transitions cause the protein to become frozen into a broad distribution of conformational states. Significant changes in the amide I region of the dehydrated proteins were observed using Fourier transform infrared spectroscopy [30-33]. Khurgin et al. [34] measured the chymotrypsin-catalyzed hydrolysis of the amide substrate N-succinyl-L-phenylalanine-p-nitroanilide at low and medium hydration levels. No enzymatic activity was observed at $w_1<0.05$ (Figure 3 from Ref. 13).

At $w_1<0.05$, the $w_2\overline{H}_2^E$ values are close to zero (Figure 1). At $w_1\sim0.05$, the $w_1\overline{H}_1^E$ functions reach maximal values. These facts imply that, at the lowest water content, changes in the excess thermodynamic functions solely reflect water addition.

Hutchens et al. [35] examined the heat capacities of insulin at w_1 = 0 and 0.038 and chymotrypsinogen A at w_1 = 0 and 0.096, from -263 to 37 °C. No phase transition corresponding to the ice-liquid water transition was observed at low hydrations.

This region corresponds to the hydration of charged groups [1,2,36]. At low water content, the $\overline{H}_1^E$ values differ significantly from zero (Figure 2). The $\overline{H}_1^E$ values are highly negative. As concluded in Ref. 13, the $\overline{H}_1^E$ (w_1=0) values are close to those observed for diethylamine, suggesting that, at the lowest w_1 values, the interaction with the protein elements containing the amino groups might be a dominant factor controlling the state of water molecules.

Part 2 (w_1=0.05-0.25) corresponds to the hydration of polar groups [1,2,36]. In this water content range, the $\overline{H}_1^E$ values change sharply from highly negative to moderate values (Figure 2). This sharp transition reflects the formation of a spanning hydrogen-bonded network of water at the protein surface [2,5]. The formation of this network occurs via a quasi two-dimensional percolation transition of the hydration water at the enzyme surface [5].

The results obtained for biopolymers using several experimental methods have been summarized in Ref. 1. The results suggest that proteins undergo a glasslike dynamic transition at water content of approximately 10 mass% at 25 °C. This water content falls within part 2 in the present study. The transition from the glassy (rigid) to the flexible (elastic) state is accompanied by significant changes in the properties of the proteins [1]. For example, the apparent heat capacity of lysozyme, ϕC_{p2}, determined from isothermal experiments using a drop calorimeter, increases from extremely low values to high values in this water content interval [37]. During the isothermal sorption of water, a glasslike transition results in a step on the excess partial enthalpy of ovalbumin (Figure 3). The $\overline{H}_2^E$ values change sharply from extremely low values to highly negative values.

Part 3 (w_1 = 0.25-0.5) indicates the appearance of the "structured" water [36]. "Structured" water is composed of molecules that interact with hydrophobic surface patches on the protein-bridging between bound water molecules. Yang and Careri [37] asserted that this region contains condensed water molecules over weakly-interacting surface elements, which are likely nonpolar atoms, not polar groups or adjacent-charged.

The onset of biological activity was observed in this region. Rupley and Careri [2] examined the lysozyme-catalyzed hydrolysis of the hexasaccharide of N-acetylglucosamine [$(GlcNAc)_6$] (Figure 3 from Ref. 13) as a function of water content and showed that the reaction grows sharply at $w_1 > 0.15$. Enzymatic activity was suppressed within regimes 1 and 2 (Figure 3 from Ref. 13).

At the highest water content $(w_1 > 0.5)$, the proteins are in a flexible (elastic) state [1]. Excess partial quantities attain their fully-hydrated values. The $\overline{H}_2^E$ values (Figure 3) reach minimal values. At $w_1 < 0.5$, the $w_1\overline{H}_1^E$ values are near zero (Figure 1).

Bulk water was observed in this region through proton NMR measurements [36]. The molecular motion of these water molecules is solely determined through the interaction characteristics of water molecules. Water addition has no significant effect on the excess thermodynamic functions. At $w_1 > 0.5$, the changes in the excess thermodynamic functions primarily reflect changes in the state of RNase A.

To show the predictive ability of the proposed methodology, we related the obtained results to general properties of the proteins. Hydrophobicity and polarity were selected as possible common parameters. The hydrophobicity of the studied proteins has been estimated as the average hydrophobicity of the side chains of the constitutive amino acids, H_ϕ (J per residue). The calculated H_ϕ values for many proteins have been collected in [38].

We calculated the areas under the $w_1\overline{H}_1^E = f(w_1)$ and $w_2\overline{H}_2^E = f(w_1)$ curves for ovalbumin (this study), lysozyme, chymotrypsinogen A, α-chymotrypsin, β-lactoglobulin, and serum albumin [13]. These areas were defined as $A(w_1\overline{H}_1^E)$ and $A(w_2\overline{H}_2^E)$, respectively. The $A(w_1\overline{H}_1^E)$ area for ovalbumin is shown shaded in Figure 1.

The $A(w_2\overline{H}_2^E)$ values are plotted as a function of H_ϕ in Figure 4. There is a clear correlation between the $A(w_2\overline{H}_2^E)$ values and H_ϕ. This correlation shows that the more hydrophobic a protein is the more significant the $A(w_2\overline{H}_2^E)$ value is.

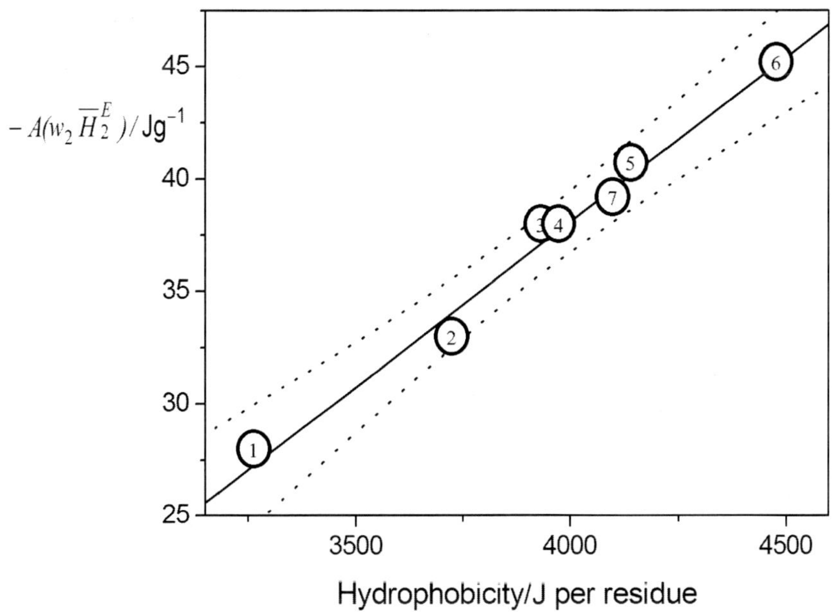

Figure 4. (1) RNase A; (2) Lysozyme; (3); α-Chymotrypsin; (4) Chymotrypsinogen A; (5) Serum albumin; (6) β-Lactoglobulin; (7) Ovalbumin.
Relationship between the $A(w_2\overline{H}_2^E)$ value and the hydrophobicity, H_ϕ, of proteins.

$-A(w_2\overline{H}_2^E) = -19.8(3.1) + 0.015(0.001) * H_\phi$, where the number of the experimental points is N=6, the standard error of estimation is S_o=0.7, and the correlation coefficient is R=0.993. The dashed lines show the 95% confidence interval.

A polarity parameter, ***P***, was proposed by Bigelow [39]. It was defined as a volume ratio of polar amino acids to nonpolar amino acids. There is a good correlation between the $A(w_1\overline{H}_1^E)$ values and Bigelow's polarity parameter (Figure 5). This correlation means that the more hydrophilic a protein is the more significant the $A(w_1\overline{H}_1^E)$ value is.

As a rule, most hydrophobic groups of globular proteins are buried in the interior of the protein molecule. On the other hand, charged and polar residues are located preferentially on the surface. H_ϕ may be regarded as a parameter characteristic mostly of the hydrophobic interior of a protein. ***P*** reflects the degree of the polarity of the protein surface.

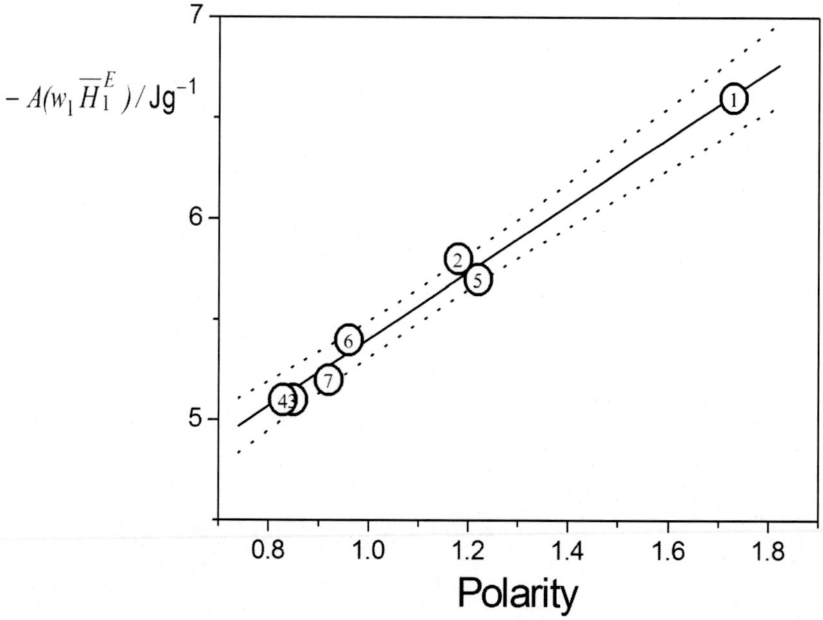

Figure 5. (1) RNase A; (2) Lysozyme; (3); α-Chymotrypsin; (4) Chymotrypsinogen A; (5) Serum albumin; (6) β-Lactoglobulin; (7) Ovalbumin. Relationship between the $A(w_1\overline{H}_1^E)$ value and Bigelow's polarity parameter, *P*, of proteins.

$-A(w_1\overline{H}_1^E) = 3.7(0.1) + 1.7(0.1) * P$, where the number of the experimental points is N=7, the standard error of estimation is S_o=0.1, and the correlation coefficient is R=0.993. The dashed lines show the 95% confidence interval.

RNase A is the most hydrophilic protein under the study. As shown in Figure 5, this protein has the most significant hydrophilic hydration contribution, $A(w_1\overline{H}_1^E)$. Lactoglobulin is the most hydrophobic protein. As shown in Figure 4, this protein has the most significant hydrophobic hydration contribution, $A(w_2\overline{H}_2^E)$. Ovalbumin shows intermediate properties.

These correlations show that the contributions corresponding to the hydration of the protein interior and the protein surface are coupled differently to the excess functions. The changes in the partial quantities corresponding to the protein interior are largely described by the $A(w_2\overline{H}_2^E)$ values. The changes in the partial quantities corresponding to the protein surface are described by the $A(w_1\overline{H}_1^E)$ values.

REFERENCES

[1] Gregory, R.B. (1995) Protein hydration and glass transition behavior. In: Protein-Solvent Interactions (Gregory, R.B., ed.), Marcel Dekker, New York, 191-264.

[2] Rupley, J.A., Careri, G. (1991) Protein hydration and function. *Adv. Protein Chem.,* 41, 37-172.

[3] Kuntz, I.D., Kauzmann, W. (1974) Hydration of proteins and polypeptides. *Adv. Protein Chem.,* 28, 239-345.

[4] Bagchi B. (2005) Water dynamics in the hydration layer around proteins and micelles. *Chem. Rev.,* 105, 3197-3219.

[5] Oleinikova, A., Smolin, N., Brovchenko, I., Geiger, A., Winter, R. (2005) Properties of spanning water networks at protein surfaces. *J. Phys. Chem. B*, 109, 1988-1998.

[6] Sirotkin, V.A., Winter, R. (2010) Volume changes associated with guanidine hydrochloride, temperature, and ethanol induced unfolding of lysozyme. *J. Phys. Chem. B*, 114, 16881-16886.

[7] Durchschlag, H., Zipper, P. (2001) Comparative investigations of biopolymer hydration by physicochemical and modeling techniques. *Biophys. Chem.*, 93, 141-157.

[8] Amberg C.H. (1957) Heats of adsorption of water vapor on bovine serum albumin. *J. Amer. Chem. Soc.,* 79, 3980-3984.

[9] Almog R., Schrier E.E. (1978) The dependence of the heat of solution of lyophilized solid ribonuclease A on water content. *J. Phys. Chem.,* 82, 1701-1702.

[10] Smith A.L., Shirazi H.M., Mulligan S.R. (2002) Water sorption isotherms and enthalpies by lysozyme using the quartz crystal microbalance/heat conduction calorimeter. *Biochim. Biophys. Acta,* 1594, 150-159.

[11] Kocherbitov V., Arnebrant T., Söderman O. (2004) Lysozyme-water interactions studied by sorption calorimetry. *J. Phys. Chem. B,* 108, 19036-19042.

[12] Sirotkin V.A., Korolev D.V. (2005) Effect of acetonitrile on the hydration of human serum albumin films: a calorimetric and spectroscopic study. *Thermochim. Acta,* 432, 246-253.

[13] Sirotkin, V.A., Khadiullina, A.V. (2011) Hydration of proteins: excess partial enthalpies of water and proteins. *J. Phys. Chem. B*, 115, 15110-15118.

[14] Sirotkin, V.A., Komissarov, I.A., Khadiullina, A.V. (2012) Hydration of proteins: excess partial volumes of water and proteins, *J. Phys. Chem. B*, 116, 4098-4105.

[15] Sirotkin, V.A., Khadiullina, A.V. (2014) A study of the hydration of ribonuclease A using densitometry: Effect of the protein hydrophobicity and polarity. *Chem. Phys. Lett.,* 603, 13-17.

[16] Sirotkin, V.A., Khadiullina, A.V. (2013) Gibbs energies, enthalpies, and entropies of water and lysozyme at the inner edge of excess hydration. *J. Chem. Phys.,* 139, 075102/1-075102/9.

[17] Yang, Y., Barendregt, A., Kamerling, J.P., Heck, A.J.R. (2013) Analyzing protein micro-heterogeneity in chicken ovalbumin by high-resolution native mass spectrometry exposes qualitatively and semi-quantitatively 59 proteoforms. *Anal. Chem.,* 85, 12037-12045.

[18] Nisbet, A. D., Saundry, R. H., Moir, A. J. G., Fothergill, L. A., Fothergill, J. E. (1981) The complete amino-acid sequence of hen ovalbumin. *Eur. J. Biochem.,* 115, 335-345.

[19] Huntington, J.A., Stein, P.E. (2001) Structure and properties of ovalbumin. *J. Chromatog. B*, 756, 189-198.

[20] Atkins, P.W. (2006) Physical Chemistry. 8th ed. Oxford: Oxford University Press.

[21] Prausnitz, J.M. (1969) Molecular Thermodynamics of Fluid-Phase Equilibria. N.J.: Prentice-Hall, Inc., Engelwood Cliffs.

[22] Belousov, V.P., Panov, M.Y. (1994) Thermodynamic properties of aqueous solutions of organic substances. Boca Raton, Fla.: CRC Press.

[23] Borisover, M.D., Sirotkin, V.A., Zakharychev, D.V., Solomonov, B.N. (2001) Calorimetric methods in evaluating hydration and solvation of solid proteins immersed in organic solvents, In: Enzymes in Nonaqueous Solvents (Vulfson, E.N., Halling, P.J., editors), Humana Press, Totowa, 183-202.

[24] Nikolskii, B.P. (1963) Spravochnik Khimika (Chemist's Handbook) Leningrad, Russia: Goskhimizdat.

[25] Greenspan, L. (1977) Humidity fixed points of binary saturated aqueous solutions. *J. Res. Natl. Bur. Stand.,* Sect. A 81, 89.

[26] Borisover, M.D., Sirotkin, V.A., Solomonov, B.N. (1995) Thermodynamics of water binding by human serum albumin suspended in acetonitrile. *Thermochim. Acta,* 254, 47-53.

[27] Sirotkin, V.A., Zinatullin, A.N. Solomonov,B.N., Faizullin, D.A., Fedotov, V.D. (2002) Interaction enthalpies of solid bovine pancreatic chymotrypsin with organic solvents: comparison with FTIR-spectroscopic data. *Thermochim. Acta,* 382, 151-160.

[28] Armstrong GT, Garvin D, Kilday MV. (1981) Standard Reference Material 1655, Potassium Chloride KCl(cr) for Solution Calorimetry. Washington, DC: National Bureau of Standards Certificate.

[29] Sabbah R, An XW, Chickos JS, Leitao MLP, Roux MV, Torres LA. (1999) Reference materials for calorimetry and differential thermal analysis. *Thermochim. Acta,* 331, 93-204.

[30] Prestrelsky S.J., Tedeschi N., Arakawa T., Carpenter J.F. (1993) Dehydration-induced conformational transitions in proteins and their inhibition by stabilizers. *Biophys. J.,* 65, 661-671.

[31] Van de Weert M., Harvez P.I., Hennink W.E., Crommelin D.J.A. (2001) Fourier transform infrared spectrometric analysis of protein conformation: effect of sampling and stress factors. *Anal. Biochem.,* 297, 160-169.

[32] Sirotkin V.A., Solomonov B.N., Faizullin D.A., Fedotov V.D.(2002) Sorption of water and acetonitrile by human serum albumin. *Russ. J. Phys. Chem. A,* 76, 2051-2057.

[33] Griebenow K., Klibanov A.M. (1995) Lyophilization-induced reversible changes in the secondary structure of proteins. *Proc. Natl. Acad. Sci. USA,* 92, 10969-10976.

[34] Khurgin YI, Medvedeva NV, Roslyakov VY. (1977) The study of solid-state enzyme reactions. II. Chymotryptic hydrolysis of N-succinyl-L-phenylalanine-p-nitroanilide. *Biofizika*, 22, 1010–1014.

[35] Hutchens, J.O., Cole, A.G., Stout, J.W. (1969) Heat capacities from 11 to 305 K and entropies of hydrated and anhydrous bovine zinc insulin and bovine chymotrypsinogen A. *J. Biol. Chem.,* 244, 26-32.

[36] Fullerton, G.D., Ord, V.A., Cameron, I.L. (1986) An evaluation of the hydration of lysozyme by an NMR titration method. *Biochem. Biophys. Acta,* 869, 230-246.

[37] Yang, P.-H., Rupley, J.A. (1979) Protein-water interactions. Heat capacity of the lysozyme-water system. *Biochemistry*, 18, 2654-2661.

[38] Bigelow CC, Channon M. (1976) *Handbook of Biochemistry and Molecular Biology,* Proteins. Cleveland, OH: CRC Press.
[39] Bigelow CC. (1967) On the average hydrophobicity of proteins and the relation between it and protein structure. *J. Theor. Biol.,* 16, 187-211.

In: Protein – Water Interactions
Editor: Vladimir A. Sirotkin
ISBN: 978-1-63463-007-8
© 2014 Nova Science Publishers, Inc.

Chapter 5

A STUDY OF THE HEAT CAPACITY OF RIBONUCLEASE A – WATER MIXTURES

Vladimir A. Sirotkin*
Kazan Federal University, A.M. Butlerov Chemical Institute, Kazan, Russia

ABSTRACT

Excess heat capacities of the binary system of bovine pancreatic ribonuclease A (RNase A) with water were obtained as a function of composition at 25°C. Differential scanning calorimetry was applied to study hydration dependencies of the excess thermodynamic functions. A major focus of this study aims to show how these thermodynamic quantities correlate with coverage of the protein by the water molecules.

The excess partial quantities are found to be sensitive to changes in the water and protein states. At the lowest water weight fractions (w_1), the changes of the excess functions can mainly be attributed to water addition. A transition from the glassy to the flexible state of the protein is accompanied by significant changes in the excess partial quantities of water and lysozyme. This transition appears at $w_1 >$ of 0.05 when charged groups of the protein are covered. Excess partial quantities reach their fully hydrated values at $w_1 > 0.5$ when coverage of both polar and weakly interacting surface elements is complete. At the highest water contents, water addition has no significant effect on the excess quantities. At $w_1 >$

* Email: vsir@mail.ru.

0.5, changes in the excess functions can solely be attributed to changes in the state of the protein.

Keywords: Hydration of proteins (biomacromolecules), Excess partial quantities, Biocalorimetry, Heat capacity

1. Introduction

Heat capacity is one of the most important thermochemical characteristics of the water-protein interactions. Heat capacity characterizes the amount of heat required to change a substance's temperature by a given amount. Below a short review of the available studies on the heat capacities of the water-protein systems. More comprehensive reviews are given in [1-3].

There are two groups of the experimental and theoretical studies. The first group describes the temperature dependencies of the heat capacity of the proteins at a fixed hydration level: (i) in the aqueous solutions [4-6] or (ii) in the dried solid state [7-10]. The second group describes the effect of hydration on the heat capacities of proteins at a fixed temperature [11-13]. A major focus of this study aims to find simultaneously the heat capacities of the protein and water at room temperature and show how these quantities correlate with coverage of the protein by the water molecules.

It was shown that the absolute values of the heat capacities of the proteins in the native state (diluted aqueous solutions) depend linearly on the temperature below the heat denaturation transition [4-6]. It was concluded that the partial heat capacity of native globular proteins varies between 1.20 and 1.50 J $g^{-1}K^{-1}$ [4-6].

The heat capacities of the dried proteins are usually lower than the heat capacities of the native proteins dissolved in water [7,11-13]. Yang and Rupley [11] measured the heat capacities of the water-lysozyme mixtures. From these experimental results they calculated the apparent heat capacity of lysozyme as a function of water content. They identified four stages in the hydration process. Stage I (0-0.07 g water g^{-1} protein) corresponds to hydration of charged groups. Stage II (0.07-0.25 g g^{-1}) corresponds to the saturation of the remaining polar sites probably associated with formation of clusters of water molecules. Stage III (0.27-0.38 g g^{-1}) represents the condensation of water over weakly interacting surface elements. Stage IV (dilute solution to 0.38 g g^{-1} to dilute solution) corresponds to the addition of water to the fully hydrated protein.

Apparent heat capacity of collagen, myoglobin, and ribonuclease was estimated in the entire range of water content, assuming Cp_1(partial water heat capacity) = 4.184 J $g^{-1}K^{-1}$ as for free and bound water [12, 13]. The heat capacities of the water-BPTI (bovine pancreatic trypsin inhibitor) mixtures were studied by molecular dynamics simulation and normal-mode analysis [14]. Apparent heat capacities of BPTI [14] were calculated in the water content range from the dried protein to the fully hydrated limit. However, the apparent heat capacity of water was only estimated for dry and hydrated proteins [11, 14]. To the best of our knowledge, no attempt has been undertaken to simultaneously estimate the protein and water contributions to the heat capacity of the binary protein-water systems in the entire range of water content.

A novel methodology has recently been proposed for studying the partial quantities of water and the proteins [15-17]. This methodology is based on the analysis of the thermodynamic functions of mixing. In this study we applied this methodology for the simultaneous monitoring of the excess partial heat capacities of water and the protein in the entire range of water content.

Bovine pancreatic ribonuclease A (RNase) was used as a model protein. This enzyme is one of the most well studied and applied in biotechnology and enzymology [18-20]. Ribonuclease A is a small monomeric protein of 124 amino acids, containing no non-enzyme components. Ribonuclease is an example of an α+β protein. RNase hydrolyzes single-stranded RNA through cleavage of the phosphodiester linkage at the 3' side of pyrimidine nucleotides [18-20].

2. Methodology

The thermodynamic properties of a real binary water-protein system can be expressed in terms of the excess functions, Z^E, i.e., the difference between the observed thermodynamic function of mixing, Z^m, and the function for an ideal binary mixture, Z^m_{id} :

$$Z^E = Z^m - Z^m_{id} \tag{1}$$

where Z^m is the thermodynamic function of mixing for a real system; Z_{id}^m is the thermodynamic function for an ideal mixture.

For binary mixtures, the Z_{id}^m values can be calculated using Eq. 2:

$$Z_{id}^m = w_1 Z_1^0 + w_2 Z_2^0 \tag{2}$$

where Z_1^0 and Z_2^0 are the thermodynamic function values for pure water and pure protein, and w_1 and w_2 are the mass fractions of water and protein, respectively ($w_2 = 1 - w_1$).

For binary mixtures, the Z^E values are composed of two components:

$$Z^E = w_1 \overline{Z}_1^E + w_2 \overline{Z}_2^E \tag{3}$$

where $\overline{Z}_1^E$ is the excess function for component 1 (water), and $\overline{Z}_2^E$ is the excess function for component 2 (protein).

The excess partial quantities of components 1 and 2 are calculated using Eqs. 4 and 5:

$$\overline{Z}_1^E = Z^E - w_2 \left(\frac{dZ^E}{dw_2}\right)_{T,P} \tag{4}$$

$$\overline{Z}_2^E = Z^E - w_1 \left(\frac{dZ^E}{dw_1}\right)_{T,P} \tag{5}$$

where $\overline{Z}_1^E$ and $\overline{Z}_2^E$ are the excess partial quantities of components 1 and 2.

The excess partial heat capacities at 25 °C and atmospheric pressure can be calculated using Eqs. 6 and 7:

$$\overline{C}_{P1}^{E} = C_P^E - w_2\left(\frac{dC_P^E}{dw_2}\right)_{T,P} \tag{6}$$

$$\overline{C}_{P2}^{E} = C_P^E - w_1\left(\frac{dC_P^E}{dw_1}\right)_{T,P} \tag{7}$$

where $\overline{C}_{P1}^{E}$ and $\overline{C}_{P2}^{E}$ are the excess partial heat capacities of water (J g^{-1} water K^{-1}) and protein (J g^{-1} protein K^{-1}), respectively.

3. EXPERIMENTAL

3.1. Materials

Bovine pancreatic ribonuclease A of the highest commercially-available purity was purchased from Sigma Chemical (St. Louis, MO) and used without further purification. The purity of the protein samples was verified through electrophoresis and dynamic light-scattering measurements (90Plus Particle Size Analyzer, Brookhaven Instruments Corporation, USA) to be more than 95%. The molecular weight of the protein was taken as 13,700 Da. The water used was doubly-distilled.

3.2. Calorimetry

All the experiments were performed using a Setaram DSC 111 instrument as described previously [21]. Heat capacity measurements of the protein-water systems were determined using the Setaram DSC 111 scanning 1° at heating rate 1° h^{-1}. Experiments were performed in a tightly closed titanium cell with a Teflon liner. The total volume of this cell was 120 µl. Typically, 20-25 mg of the dried or hydrated protein sample was placed in the sample titanium cell. Before and after scanning, a steady baseline was obtained and the heat capacity was determined by the integration of the excess heat flow over the temperature interval. The instrument was calibrated for heat capacity measurements with a sample of synthetic sapphire (C_P =0.775 J g^{-1} K^{-1} at 298

K [22]). Temperature calibration was achieved using the melting of indium (T_m 429.6 K) and naphthyl ethyl ester (T_m 300.5 K). The heat capacity was determined to better than 1%.

3.3. Water Sorption Measurements

The dried protein preparation (zero hydration level) was obtained through vacuum drying using a microthermoanalyzer "Setaram" MGDTD-17S (±0.00001 g) at 25 °C and 0.1 Pa, until a constant sample mass was reached. The water content of the dried protein was estimated as 0.003 ± 0.002 g water g^{-1} protein using the Karl Fischer titration method, according to the recommendations [21]. This zero hydration level was described using $\overline{H}_P^0$ (Eqs. 4-6).

At the lowest and intermediate water mass fractions, the protein preparations were prepared as follows. The initially dehydrated protein samples were exposed to pure water vapor. The water vapor was consecutively flowed through a thermostated glass tube with drying agent (P_2O_5), a thermostated saturator filled with saturated salt solution, and a cell containing the protein sample.

The schematic representation of the experimental setup is given in Figure 1 from Ref 23. An external ethylene glycol thermostat (RC 6 from Lauda, Germany) was used to control the temperature with a precision of 0.1 °C. Sorption equilibrium was reached after 180 min. The water activity (a_w) in the vapor phase was adjusted by changing the saturated salt solution in the saturator. The water activities of the saturated salt and KOH solutions were taken from [24, 25].

The following substances were used: LiBr (a_w=0.064), KOH (a_w=0.078), LiCl (a_w=0.11), $CaBr_2$ (a_w=0.17), CH_3COOK (a_w=0.22), $MgCl_2$ (a_w=0.33), K_2CO_3 (a_w=0.44), $Mg(NO_3)_2$ (a_w=0.53), NaCl (a_w=0.75), KCl (a_w=0.84), KNO_3 (a_w=0.94), and K_2SO_4 (a_w=0.97). The salts used for sample conditioning were of analytical pure grade.

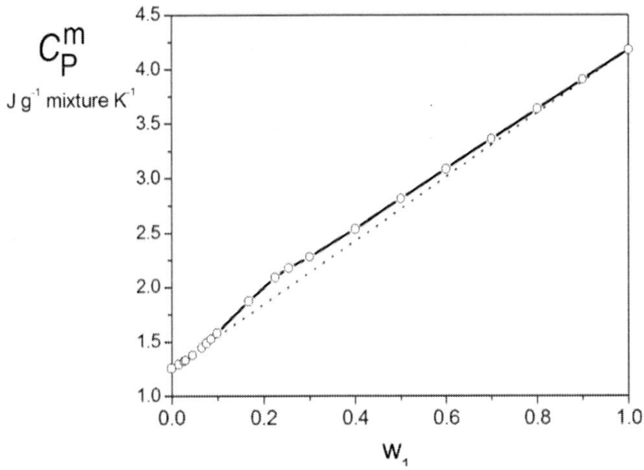

Figure 1. The C_P^m value for the water-RNase A mixtures as a function of the mass fraction of water, w_1, at 25 °C. The standard errors of estimation of the C_P^m values were 0.01-0.02 J g^{-1}K^{-1}. Each experiment was performed 3-4 times. The dashed line corresponds to the ideal water-protein mixture.

The water content of the samples after equilibration was measured by drying under vacuum using a microthermoanalyzer "Setaram" MGDTD-17S at 25 °C and 0.1 Pa, until a constant sample mass was reached.

At the highest water mass fractions (a_w>0.98), the water activity was adjusted by changing the unsaturated NaCl-water solutions at 25 °C. The a_w values of the NaCl-water solutions were taken from Ref. 28. This technique was used for the protein-water systems for $w_1 > 0.6$.

4. Results and Discussion

Figure 1 shows the C_P^m values in the entire range of water content. At 25 °C, the heat capacities values of the dried RNase A and pure liquid water are: 1.26±0.02 J g-1 K-1 (or 17.26 J mol^{-1} K^{-1}) and 4.183±0.01 J g^{-1} K^{-1} (or 75.3 J mol^{-1} K^{-1}). These values are in agreement with the previously published results for RNase A (17.23 J g^{-1} K^{-1} [9]) and water (75.3 J mol^{-1} K^{-1} [26] indicating that the apparatus and the experimental procedure are adequate to calculate the heat capacities values.

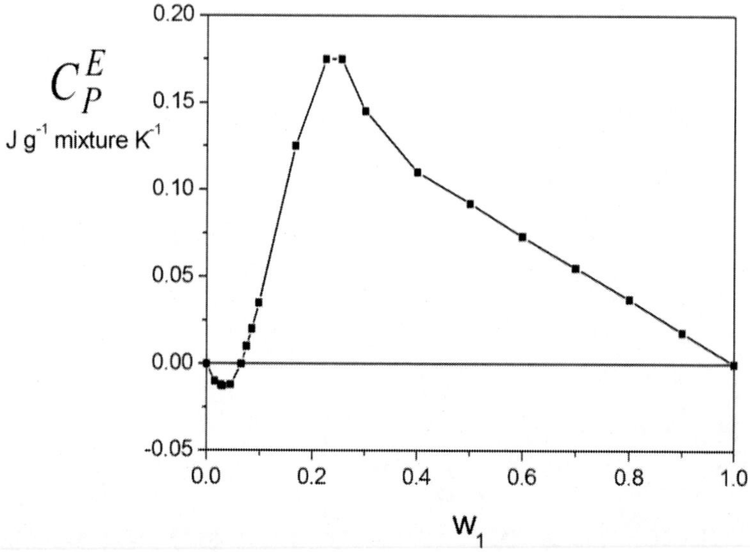

Figure 2. The C_P^E value as a function of the mass fraction of water, w_1, at 25 °C.

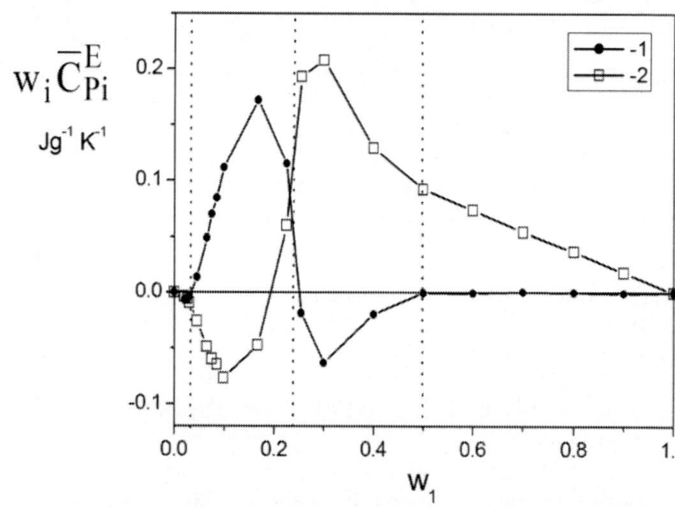

Figure 3. Excess thermodynamic functions plotted versus the mass fraction of water, w_1, at 25 °C: (1) $w_1 \overline{C}_{P1}^E$; (2) $w_2 \overline{C}_{P2}^E$.

Figure 2 shows how the C_P^E values depend on the water weight fraction. The C_P^E values were calculated using Eq. (1).

In ideal binary mixtures (mixtures of two components, W [water] and P [protein]) the average W-P interactions are the same as the average W-W and P-P interactions. Non-ideal mixtures are composed of particles for which the W-W, P-P, and W-P interactions are all different. As shown in Figure 2, the C_P^E values differ significantly from zero, indicating that the water-protein mixtures are non-ideal in the entire range of water content. Figure 3 present the dependencies of the $w_i\overline{Z}_i^E$ ($w_1\overline{C}_{P1}^E$ and $w_2\overline{C}_{P2}^E$) functions plotted versus the water mass fraction. These functions were calculated using Eq. 3. They show the individual protein and water contributions to the excess thermodynamic functions in the entire range of water content. Figure 3 demonstrates that the protein contribution ($w_2\overline{C}_{P2}^E$) reaches maximal values at w_1~0.25. However, the $w_1\overline{C}_{P1}^E$ function reaches maximal values at low water content (w_1~0.05).

Figures. 4 and 5 present the partial heat capacity of water, $\overline{C}_{P1}^E$, and RNase A, $\overline{C}_{P2}^E$. These thermochemical quantities were calculated using Eqs. 6 and 7. The partial heat capacity curves presented in Figures. 4 and 5 can be divided into four parts.

4.1. Regime 1. (w_1=0-0.05).

The main features of regime 1 can be described as follows. This region corresponds to the hydration of charged groups [1,2,27]. At the lowest water content, the proteins are in a glassy (rigid) state [1]. In the glassy state, the dehydration-induced conformational changes and restrictions on conformational transitions cause the protein to become frozen into a broad distribution of conformational states. No biological activity was observed at the lowest water content (Figure 3 from Ref. 15).

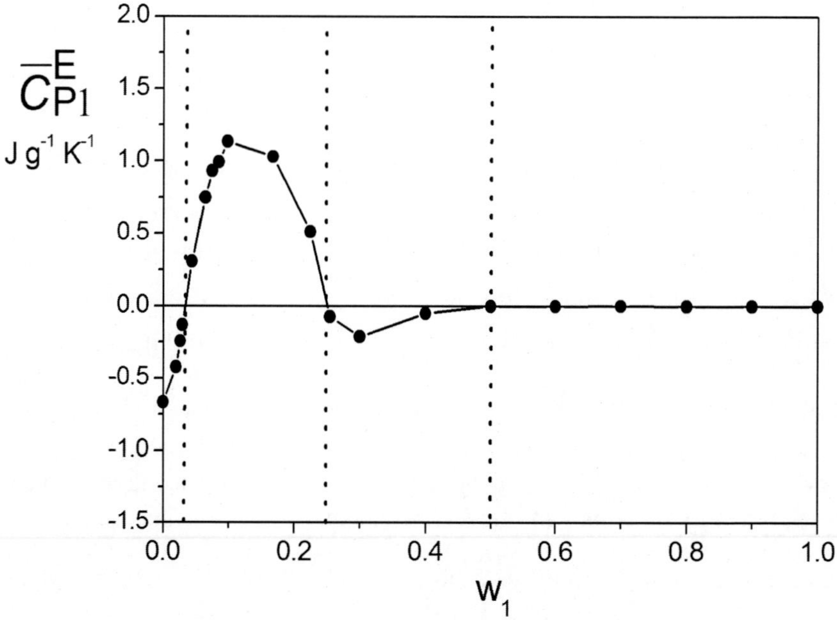

Figure 4. Excess partial heat capacity of water, $\overline{C}^{E}_{P1}$, as a function of the mass fraction of water, w_1, at 25°C.

At the lowest water content, the excess partial heat capacity of RNase A, $\overline{C}^{E}_{P2}$, is close to zero and do not depend noticeably on the water content (Figure 4). The fact that RNase A is in a glassy state explains this feature of Regime 1. The $\overline{C}^{E}_{P2}$ values are rather constant due to the fact that all the protein molecules came into contact mainly with the same protein molecules during this range of water contents.

At w_1<0.05, the $w_2\overline{C}^{E}_{P2}$ values are close to zero (Figure 3). The excess partial heat capacity of water, $\overline{C}^{E}_{P1}$, is negative (Figure 4). At w_1~0.05, the $w_1\overline{C}^{E}_{P1}$ function reaches maximal values. These facts imply that changes in the excess functions (Figure 3) can solely be attributed to water addition.

4.2. Regime 2 (w_1=0.05-0.25)

Regime 2 corresponds to hydration of polar groups [1,2,27]. In the water content range $w_1=0.05\text{-}0.25$, the $\overline{C}^E_{P2}$ values change moderately from zero to negative (~-0.1 J g^{-1} K^{-1}) values (Figure 5). The $\overline{C}^E_{P1}$ values increase sharply from negative (~-0.2 J g^{-1} K^{-1}) to positive (~1.8 J g^{-1} K^{-1}) values. These changes are in agreement with those observed for the water-lysozyme mixtures [11].

The partial heat capacity of water is higher than the heat capacity of pure water within regime 2 (Figure 4). There is a maximum on the $\overline{C}^E_{P1} - w_1$ curve. Considerable change in the $\overline{C}^E_{P1}$ values within regime 2 was attributed to the formation of associated forms of the hydration water. Probably, there is a relative increase of the proportion of the high-density domains of water in the vicinity of hydrophilic protein groups.

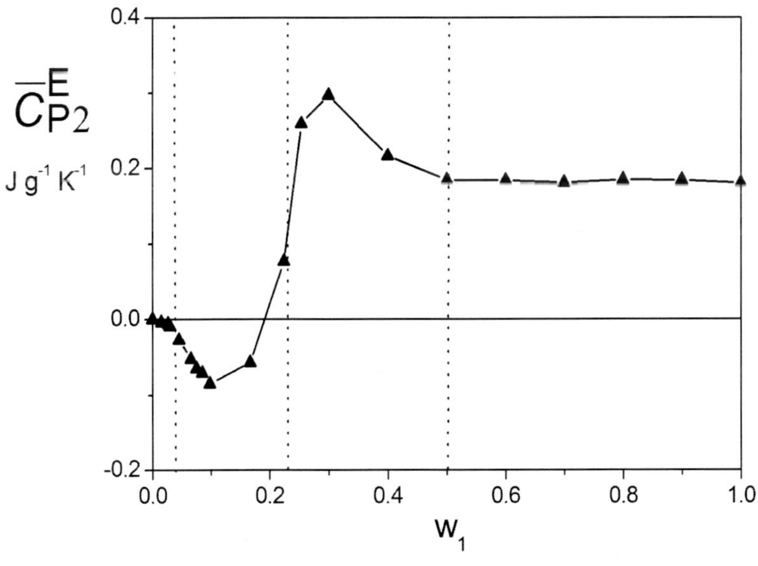

Figure 5. Excess partial heat capacity of RNase A, $\overline{C}^E_{P2}$, as a function of the mass fraction of water, w_1, at 25°C.

The $\overline{C}_{P2}^{E} - w_1$ curve goes through a minimum (Figure 5). This means that the partial heat capacity of lysozyme within regime 2 is lower than the partial heat capacity of RNase A in the diluted aqueous solution. The existence of the minimum on the $\overline{C}_{P2}^{E} - w_1$ curve was explained by the hydrophilic hydration of polar groups of ribonuclease A.

4.3. Regime 3 (w_1=0.25-0.5).

Regime 3 corresponds to the appearance of the "structured" water [27]. It was concluded that the "structured" water consists of molecules which interact with hydrophobic surface patches on the protein while bridging between bound water molecules. Rupley and Careri [2] attributed this region to the condensation of water molecules over weakly interacting surface elements, probably nonpolar atoms not adjacent charged and polar groups. The $\overline{C}_{P2}^{E} - w_1$ curve goes through a maximum (Figure 5). This means that the partial heat capacity of RNase A within regime 3 is higher than the partial heat capacity of RNase A in pure liquid water. The existence of the maximum in the $\overline{C}_{P2}^{E} - w_1$ curve was explained by the hydrophobic hydration of nonpolar surface regions of RNase A. A similar effect was observed in water-rich mixtures of various organic solvents containing hydrophobic groups [28,29].

The partial heat capacity of water is lower than the heat capacity of pure water within regime 3 (Figure 4). The existence of the minimum on the $\overline{C}_{P1}^{E} - w_1$ curve was explained by a relative increase of the proportion of the bulky, low-density, domains of water in the vicinity of hydrophobic protein regions.

The onset of the biological activity was observed in this region. Khurgin et al. [30] measured the α-chymotrypsin-catalyzed hydrolysis of the amide substrate N-succinyl-L-phenylalanine-p-nitroanilide at low and medium hydration levels. Biological activity is suppressed within regimes 1 and 2 (Figure 3 from Ref. 15). As concluded from Figure 3 from Ref. 15, the reaction grows sharply at w_1>0.15.

The most important observation is the correlations presented in Figure 3-5. These correlations show that the contributions corresponding to the protein interior and the protein surface are coupled differently to the excess functions. The changes in the excess partial heat capacities corresponding to the hydration of the protein hydrophilic regions are largely complete within regime 2. The changes in the excess partial heat capacities corresponding to the hydration of the protein hydrophobic regions are complete within regime 3.

4.4. Regime 4. ($w_1 > 0.5$).

The main features of regime 4 can be described as follows. At high water content, the proteins are in a flexible (elastic) state [1]. Excess partial quantities reach their fully hydrated values at water weight fraction more than 0.5 when coverage of both polar and adjacent weakly interacting surface elements no longer changes appreciably upon further hydration.

The excess partial heat capacities, $\overline{C}_{P2}^{E}$, of RNase A reach constant values (Figure 5). The $\overline{C}_{P1}^{E}$ values are close to zero (Figure 4). Bulk water was observed in this region from proton NMR measurements [27]. Molecular motion of these water molecules is determined solely by the interaction characteristics of water molecules. This means, at $w_1>0.5$, water addition has no significant effect on the thermodynamic excess functions. The changes of the thermodynamic excess functions can mainly be attributed to changes in the state of the proteins. In other words, the protein hydration is full and water molecules added at $w_1>0.5$ are not perturbed by the protein macromolecules.

Conclusion

In the present study, differential scanning calorimetry and water sorption measurements were applied to study the hydration of ribonuclease A. The experimental data on the heat capacity of binary ribonuclease A-water systems have been determined in the entire range of water content at 298 K. The hydration process was characterized by analyzing the excess partial quantities. This methodology allows for studying separately the protein and water partial quantities in the entire range of water content.

It was shown that the protein and water contributions to the excess functions depend markedly on the hydration level. At the lowest water contents, changes of the excess functions can solely be attributed to water addition. A transition from the glassy to the flexible state of proteins is accompanied by significant changes in the partial quantities of water and the protein.

This transition appears in the calculated quantities when charged groups of proteins are covered, which occurs at a water weight fraction of 0.05 and 298 K. Partial quantities reach their fully hydrated values at a water weight fraction more than 0.5 when coverage of both polar and adjacent weakly interacting surface elements no longer changes appreciably upon further hydration. At the highest water contents, water addition has no significant effect on the excess functions. At $w_1 > 0.5$, the changes of the excess function can mainly be attributed to changes in the state of ribonuclease A.

REFERENCES

[1] Gregory, R. B. (1995). Protein hydration and glass transition behavior. In Protein-Solvent Interactions (Gregory, R. B., Ed.), pp. 191-264, Marcel Dekker: New York.

[2] Rupley, J.A., Careri, G. (1991) Protein hydration and function. *Adv. Protein Chem.,* 41, 37-172.

[3] Kuntz, I.D., Kauzmann, W. (1974) Hydration of proteins and polypeptides. *Adv. Protein Chem.,* 28, 239-345.

[4] Privalov, P.L. (1979) Stability of proteins. Small globular proteins. *Adv. Protein Chem., 33*, 167-241.

[5] Privalov, P. L., Makhatadze, G. I. (1990) Heat capacity of proteins. II. Partial molar heat capacity of the unfolded polypeptide chain of proteins: Protein unfolding effects. *J. Mol. Biol.,* 213, 385-391.

[6] Velicelebi, G., Sturtevant, J. M. (1979) Thermodynamics of the denaturation of lysozyme in alcohol–water mixtures, *Biochemistry,* 18, 1180-1186.

[7] Hutchens, J.O., Cole, A.G., Stout, J.W. (1969) Heat capacities from 11 to 305 K and entropies of hydrated and anhydrous bovine zinc insulin and bovine chymotrypsinogen A. *J. Biol. Chem.,* 244, 26-32.

[8] Bakk, A. (2002) Heat capacities of solid state proteins: implications for protein stability in solution. *Physica A*. 313, 540-548.

[9] Zhang G., Gerdes, S., Wunderlich B. (1996) Heat capacities of solid, globular proteins. *Macromol. Chem. Phys*., 197, 3791-3806.

[10] Di Lorenzo M. L., Zhang G., Pyda, M., Lebedev, B. V., Wunderlich B. (1999) Heat capacity of solid-state biopolymers by thermal analysis. *J. Polymer. Sci: Part B: Polymer Phys.,* 37, 2093-2102.

[11] Yang, P.-H., Rupley, J.A. (1979) Protein-water interactions. Heat capacity of the lysozyme-water system. *Biochemistry,* 18, 2654-2661.

[12] Tsereteli G. I., Belopolskaya T. V., Grunina N. A., Vaveliouk O.L. (2000) Calorimetric study of the glass transition process in humid proteins and DNA. *J. Therm. Anal. Cal.,* 2000, 62, 89-99.

[13] Tsereteli G. I., Belopolskaya T. V., Grunina N. A. (2008) Dehydrated native biopolymers – a unique representative of glassy systems. *J. Therm. Anal. Cal.,* 2008, 92, 711-716.

[14] Yu, X., Park, J., Leitner, D.M. (2003) Thermodynamics of protein hydration computed by molecular dynamics and normal modes. *J. Phys. Chem. B*. 107**,** 12820-12829.

[15] Sirotkin, V.A., Khadiullina, A.V. (2011) Hydration of proteins: excess partial enthalpies of water and proteins. *J. Phys. Chem. B*, 115**,** 15110-15118.

[16] Sirotkin, V.A, Komisarov, I.A., Khadiullina, A.V. (2012) Hydration of proteins: excess partial volumes of water and proteins. *J. Phys. Chem. B*, 116, 4098-4105.

[17] Sirotkin, V.A.; Khadiullina, A.V. (2011) Hydration of chymotrypsin: excess partial enthalpies of water and enzyme. *Thermochim. Acta,* 522**,** 205-210.

[18] Fersht A. Structure and mechanism in protein science: a guide to enzyme catalysis and protein folding. New York: Freeman & Co, 1999.

[19] Lehninger A.L., Nelson D.L., Cox M.M. Principles of Biochemistry. New York: Worth, 1993.

[20] Raines R.T. (1998) Ribonuclease A. *Chem. Rev.,* 98: 1045-1065.

[21] Borisover, M.D., Sirotkin, V.A., Zakharychev, D.V., Solomonov, B.N. (2001) Calorimetric methods in evaluating hydration and solvation of solid proteins immersed in organic solvents, In: Enzymes in Nonaqueous Solvents (Vulfson, E.N., Halling, P.J., editors), Humana Press, Totowa, 183-202.

[22] Hohne, G.W.H., Hemminger, W., Flammerschein, H.-J. (1996) Differential Scanning Calorimetry. An Introduction for Practitioners, Springer-Verlag: Berlin, Heidelberg, New York.

[23] Sirotkin, V.A., Khadiullina, A.V. (2013) Gibbs energies, enthalpies, and entropies of water and lysozyme at the inner edge of excess hydration. *J. Chem. Phys.,* 139, 075102/1-075102/9.

[24] Nikol'skii, B. P. (1963). Spravochnik Khimika (Chemist's Handbook); Goskhimizdat: Leningrad, Russia.

[25] Greenspan, L. (1977). Humidity fixed points of binary saturated aqueous solutions. *J Res. Natl. Bur. Stand., Sect. A,* 81, 89-96.

[26] Lide, D. R. (2007) Handbook of Chemistry and Physics, 86th edn. CRC: Boca Raton.

[27] Fullerton, G.D., Ord, V.A., Cameron, I.L. (1986) An evaluation of the hydration of lysozyme by an NMR titration method. *Biochem. Biophys. Acta,* 869, 230-246.

[28] Koga, Y., Siu, W.W.Y., Wong, T.Y.H. (1990). Excess partial molar free energies and entropies in aqueous tert-butanol solutions at 25°C. *J. Phys. Chem.,* 94, 7700 - 7706.

[29] Franks, F. (1982). Hydrophobic interactions - a historical perspective. *Faraday Symp. Chem. Soc.,* 17, 7-10.

[30] Khurgin YI, Medvedeva NV, Roslyakov VY. (1977) The study of solid-state enzyme reactions. II. Chymotryptic hydrolysis of N-succinyl-L-phenylalanine-p-nitroanilide. *Biofizika,* 22, 1010–1014.

In: Protein – Water Interactions
Editor: Vladimir A. Sirotkin

ISBN: 978-1-63463-007-8
© 2014 Nova Science Publishers, Inc.

Chapter 6

EFFECT OF WATER SOLVATION ON THE HEAT EFFECTS OF DEHYDRATION OF ALPHA-CHYMOTRYPSIN IN ORGANIC SOLVENTS

Vladimir A. Sirotkin [*] ***and Elena V. Dudkina***
Kazan Federal University, A.M. Butlerov Institute of Chemistry, Kazan, Russia

ABSTRACT

Enthalpy changes on the interaction of the dried and hydrated bovine pancreatic α-chymotrypsin (CT) with anhydrous organic solvents (dimethyl sulfoxide (DMSO), formamide, ethylene glycol, and methanol) were measured by isothermal calorimetry at 25°C. Initial hydration level of CT was varied in a wide range of water content (from 0 to 0.3 g water/g enzyme). The interaction enthalpies of the dried CT with anhydrous organic solvents are exothermic. At high water content (more than 0.2 g water/g enzyme), the interaction enthalpies are endothermic for formamide and exothermic for DMSO, methanol, and ethylene glycol.

These thermochemical data were analyzed to calculate the molar enthalpies of dehydration of the enzymes in organic liquids. The dehydration enthalpy changes may be endothermic or exothermic depending on the initial water content and the water solvation enthalpy

[*] E-mail: vsir@mail.ru (V.A. Sirotkin).

value. The most important conclusions can be described as follows: (i) The solvation of water by hydrophilic organic solvent determines the changes in the dehydration enthalpy values; (ii) The data for the enthalpies of solvation of water by the solvent at infinite dilution reflect this effect.

Keywords: Enzyme hydration, Heat effects, Calorimetry, Water solvation, Organic solvent

1. INTRODUCTION

This study aims to solve a topical problem – the elucidation of the patterns of the thermodynamics of enzymes in low water organic solvents. Given their ability to vary the size, polarity, and strength of hydrogen bonds, organic solvents are used increasingly in biophysical chemistry and biotechnology to selectively modulate the properties of enzyme-water systems. There are many advantages in employing nonaqueous media for biocatalysis, including the high solubility of hydrophobic reagents, the synthesis of useful chemicals, the suppression of undesirable side reactions caused by water, and increased thermostability [1-3]. High selectivity, including substrate and stereoselectivity, can be markedly affected, and sometimes even inverted, by the solvent. Enzymatic catalysis in organic solvents (ionic liquids, supercritical fluids) is a competitive and cost-saving technology for producing substances with high optical purity.

However, the efficiency of these biocatalysts and biosorbents depends markedly on the hydration level, but in a complicated way [4-8]. Consequently, studying the state of water at the protein surface in organic media may facilitate the development of novel efficient biocatalytic systems. In addition, such fundamental studies make it possible to extend our knowledge about the protein macromolecules and the role that water plays in enzyme functioning. Hence, there is a clear need for experimental and theoretical approaches by which the characteristics of the enzyme-water interactions in the presence of organic solvents may be determined.

Calorimetry is one of the effective methods for obtaining reliable thermochemical information on the interactions of enzymes with water in various environments, including organic solvents. Thus, particular, Smith et al. [9] calorimetrically measured the heats of water adsorption on lysozyme in the range of relative water vapor pressures from 0 to 0.895. They obtained both the sorption isotherm and the enthalpy of hydration of the protein in the water content range 0-18 % (g water/g enzyme) at 25°C. Kocherbitov et al. [10] used sorption calorimetry to measure the adsorption isotherm of water on lysozyme and the corresponding heat effects in the entire range of water activities. Our research group developed an experimental method for measuring the heat effects of hydration-dehydration of proteins in the presence of organic solvents [11-13]. The interaction enthalpy was found to depend significantly on the initial water content and hydration history.

However, similar information on the enzyme-water interactions in organic media is more limited and intricate. For example, Dordick and Gorman [14] studied the desorption of the enzyme-bound T_2O into nonaqueous medium for chymotrypsin, subtilisin Carlsberg, and horseradish peroxidase. They showed that enzymes lost the enzyme-bound T_2O in organic solvents. Polar solvents resulted in the highest degree of T_2O desorption. However, no quantitative correlation was observed between the desorption ability of organic liquids and solvent polarity, hydrophobicity, and the saturated molar solubility of water in a given solvent.

Therefore, thermochemical studies of the enzyme-water interactions may be extremely pertinent to the understanding of the functioning of enzymes in non-aqueous organic solvents. The enthalpy changes accompanying the interaction of the dried α-chymotrypsin with pure organic solvents including hydrocarbons, alcohols, and hydrogen bond accepting solvents have been studied using isothermal calorimetry [15]. It was found that the enthalpy changes depend cooperatively on the solvent hydrophilicity. The solvent hydrophilicity was characterized by the Gibbs energy of dissolution of water in an organic solvent. Based on this solvent hydrophilicity parameter, all the solvents were divided into two groups. The first group included the hydrophilic solvents, e.g., methanol and DMSO. Considerable structural rearrangements were observed in this group of solvents. The interaction enthalpies of the dried enzyme with hydrophilic liquids were strongly exothermic. The second group included the hydrophobic and medium hydrophilic liquids, e.g., benzene, dioxane, butanol-1, propanol-1. The enthalpy and structural changes in the second group of solvents were close to zero.

In the present study, we reported new data on the interaction enthalpies of the dried and hydrated enzyme with non-aqueous hydrophilic organic solvents. Initial hydration level was varied in the entire range of water contents (from 0 to 30 % (g water/g enzyme)). These thermochemical results were compared with the water sorption data to calculate the molar enthalpy of the enzyme dehydration in organic liquids. The aim of this work was to develop a thermochemical model for describing the transfer of water from the enzyme phase to the organic solvent liquid phase and for determining how the solvation ability of hydrophilic organic liquids affects this process.

Bovine pancreatic α-chymotrypsin (CT) was used as a model enzyme. It is one the most studied and applied in biochemistry, molecular biology and enzymology [16, 17]. Physiological role of CT is to hydrolyze peptide bonds [16, 17]. α-Chymotrypsin is an example of a predominantly β-sheet protein.

Hydrophilic organic solvents used were dimethyl sulfoxide (DMSO), formamide, methanol, and ethylene glycol. These water-miscible organic solvents represent a series of liquids in which the enthalpy of solvation of water is gradually changed in a wide range.

2. EXPERIMENTAL SECTION

2.1. Materials

Bovine pancreatic α-chymotrypsin (EC 3.4.21.1, Sigma, C 4129, essentially salt free) was used without further purification. The molecular weight of bovine pancreatic α-chymotrypsin was taken as 25,000 Da. Organic solvents (reagent grade, purity >99%) were purified and dried according to the recommendations [17]. Water used was double distilled.

2.2. Calorimetry

The enthalpy changes (ΔH_{tot}) on the immersion of the dried and hydrated enzyme preparations into pure liquid water or binary water – organic mixtures were measured at 25°C with a Setaram BT-2.15 calorimeter according to the described procedure [15,19]. Typically, the sample of 5-10 mg of CT contacted with 4.0 mL of a solvent in the calorimetric cell. Typical heat evolution curves recorded on the interaction of proteins with pure liquid water or organic solvents are given in Ref. [15, 19]. The heat evolution was completed for 40 min. Calorimeter was calibrated using the Joule effect and tested with dissolving sodium chloride in water according to the recommendations [19].

The dried enzyme preparations (**zero hydration level**) were obtained by drying under vacuum using a microthermoanalyzer "Setaram" MGDTD-17S at 25°C and 0.1 Pa until the constant sample weight was reached. Water content of the dried enzymes was estimated as 0.2 ± 0.1 % (g water/g enzyme) by the Karl Fischer titration method according to the recommendations [19]. Samples for the determination of the interaction enthalpies of the hydrated proteins with water were equilibrated at 25°C ±0.5°C in tightly closed desiccator over the saturated salt solutions. The following salts were used: LiCl, CH_3COOK, $MgCl_2$, K_2CO_3, $Mg(NO_3)_2$, NaCl, $BaCl_2$. Water activities over the saturated salt solutions were taken from [20, 21]. The masses of protein samples used in the equilibration were in the range of 8-10 mg. Water content of the samples after equilibration was measured by drying using a microthermoanalyzer "Setaram" MGDTD-17S at 25°C and 0.1 Pa until the constant sample weight was reached. Salts for the conditioning of the samples were of analytical pure grade. The conditioned samples were then taken from the desiccator and equilibrated in the calorimetric cell at 25°C before the experiment.

2.3. Water Content of Organic Solvents

The water content of organic solvents was determined using the Karl Fisher method. The equilibrium water concentrations in organic solvents after immersion of the enzyme preparations were lower than 0.04 mol/L for ethylene glycol, 0.05 mol/L for formamide, 0.07 mol/L for DMSO, and 0.02 mol/L for methanol. The corresponding water activities calculated using Eq.1 did not exceed 0.01 in all the solvents studied:

$$a_W = \gamma_W x_W \quad (1)$$

where x_W is the mole fraction of water in the solution and γ_W is the activity coefficient of water (in mole fractions; the standard state is pure water). Water activity coefficients (γ_W) were calculated using literature data [22] on the vapor-liquid equilibrium using Eq. 2:

$$\gamma_W = \frac{y_W P_{tot}}{x_W P_W^o} \quad (2)$$

where y_W is the measured mole fraction of water in vapor phase, P_t is the total pressure, P_o is the saturated vapor pressure of pure water at the same temperature and x_W is the mole fraction of water in the liquid phase.

2.4. Enthalpies of Solvation of Water in Organic Solvents

These thermochemical parameters were calculated using Eq. 3:

$$\Delta H_{solv}^{H_2O/S} = \Delta H_{sol}^{H_2O/S} - \Delta H_{vap}^{H_2O} \quad (3)$$

where $\Delta H_{sol}^{H_2O/S}$ is the enthalpy of solution of water in a solvent at infinite dilution and 25°C (kJ/mol); $\Delta H_{vap}^{H_2O}$ is the enthalpy of vaporization of water (43.7 kJ/mol) [20]. Consequently, the $\Delta H_{solv}^{H_2O/S}$ value for water is equal to –43.7 kJ/mol. The $\Delta H_{solv}^{H_2O/S}$ values for DMSO, methanol, ethylene glycol, and formamide were found to be -49.0, -47.1, -45.8, and -42.7 kJ/mol, respectively. The enthalpies of solution of water in organic solvents at infinite dilution and 25°C were taken from [23,24].

3. METHODOLOGY

The studied systems can be described as follows:

(1) Initially, the solid enzyme phase does not contain organic component. The initial enzyme phase may contain water.

(2) The final enzyme phase may contain organic component. The final enzyme phase does not contain water.

The interaction enthalpy of the dried or hydrated enzyme with anhydrous organic solvent, ΔH_{tot}, is given by Eq. 4:

$$\begin{aligned}
&\Delta H_{tot}(\,hydrated\,) = [\,\overline{H}_W m_w + \overline{H}_S \cdot m_S\,] \cdot_{final \cdot liquid} + \\
&[\,\overline{H}_E \cdot m_E + \overline{H}_W m_w + \overline{H}_S \cdot m_S\,]_{final \cdot enzyme} - \\
&[\,\overline{H}_W m_w + \overline{H}_S \cdot m_S\,]_{initial \cdot liquid} - \\
&[\,\overline{H}_E \cdot m_E + \overline{H}_W m_w\,]_{initial \cdot \cdot enzyme}
\end{aligned} \tag{4}$$

where $\overline{\boldsymbol{H}}_E$, $\overline{H}_W$, $\overline{H}_S$ are the partial enthalpies of the enzyme, water, and organic components, respectively; m_E, m_W, m_S are mass amounts of the enzyme, water, and organic component, respectively. Phases (liquid or solid) and states (final or initial) are specified by subscripts. The amounts of water, m_W^{tr}, and organic component, m_S^{tr}, transferred from the liquid phase to the enzyme phase are defined as in Eqs. 5 and 6:

$$\begin{aligned}
&m_W^{tr} = [\,m_{W,.final} - m_{W,.initial}\,]_{liquid} = \\
&[\,m_{W,.initial} - m_{W,.final}\,]_{enzyme}
\end{aligned} \tag{5}$$

$$\begin{aligned}
&m_S^{tr} = [\,m_{S,.final} - m_{S,.initial}\,]_{enzyme} = \\
&[\,m_{S,.initial} - m_{S,.final}\,]_{liquid} = m_{S,.final.enzyme}
\end{aligned} \tag{6}$$

When the enzyme–liquid ratio is small enough (In our work the molar ratio of enzyme/organic solvent is about $2{*}10^{-6}$), the partial enthalpies of water and organic component in the liquid phase are not changed during enzyme sample immersion. Therefore, Eq. 4 can be transformed into an expression useful for experimental applications, given by Eq. 7:

$$\begin{aligned}
&\Delta H_{tot}(\,hydrated\,) = \overline{H}_{W,\cdot liquid} m_W^{tr} + \\
&\overline{H}_W m_{W,\cdot final \cdot \cdot enzyme} - \overline{H}_W m_{W,\ initial \cdot \cdot enzyme} + \\
&m_E[\,\overline{H}_{E,\,final \cdot} - \overline{H}_{E,initial \cdot}\,] + m_S^{tr}[\,\overline{H}_{S,.enzyme} - \overline{H}_{S,\ liquid}\,]
\end{aligned} \tag{7}$$

where $\Delta H_{tot}(hydrated)$, m_{W}^{tr}, m_{S}^{tr} are related now to the unit mass amount of enzyme.

When the transfer of water from the enzyme phase to the liquid phase is complete, i.e., $m_{w,\cdot final\cdot\cdot enzyme} = 0$ and $m_{W}^{tr} = m_{w,\cdot initial\cdot\cdot enzyme}$, Eq. 7 can be transformed into Eq. 8:

$$\Delta H_{tot}^{o}(hydrated) = m_{W}^{tr}[\overline{H}_{W,\ \cdot initial\cdot liquid} - \overline{H}_{W,\ \cdot initial.enzyme}] + m_{E}[\overline{H}_{E,\ final\cdot} - \overline{H}_{E,initial\cdot\cdot}] + m_{S}^{tr}[\overline{H}_{S,.final.enzyme} - \overline{H}_{S,\ .initid..liquid}] \quad (8)$$

In the presence of water in the initial liquid phase, $m_{w,\cdot initial\cdot\cdot enzyme} = 0$, Eq. 7 can be transformed into Eq. 9:

$$\Delta \boldsymbol{H}_{tot}(dried) = \overline{\boldsymbol{H}}_{w,.liquid} m_{W}^{tr} + \overline{\boldsymbol{H}}_{W} m_{w,\cdot final\cdot\cdot enzyme} + m_{E}[\overline{\boldsymbol{H}}_{E,\ final\cdot\cdot} - \overline{\boldsymbol{H}}_{E,initial\cdot\cdot}] + m_{S}^{tr}[\overline{\boldsymbol{H}}_{S,.final.enzyme} - \overline{\boldsymbol{H}}_{S,\ .initial.liquid}] \quad (9)$$

In the absence of water in the enzyme and liquid phase, the $\Delta \boldsymbol{H}_{tot}^{o}(dried)$ values demonstrate immediately the energetics of enzyme–organic solvent interactions, Eq. 10. Depending on the specific mechanism, these enthalpy changes may be considered as wetting heats (if solvent molecules solvate the surface of the enzyme sample) or swelling heats (if the solvent molecules penetrate into the enzyme bulk).

$$\Delta H_{tot}^{o}(dried) = m_{E}[\overline{H}_{E,\ final\cdot\cdot} - \overline{H}_{E,initial\cdot\cdot}] + m_{S}^{tr}[\overline{H}_{S,\cdot final\cdot\cdot enzyme} - \overline{H}_{S,\ \cdot initial\cdot\cdot liquid}] \quad (10)$$

In order to characterize the enzyme–water interactions in the presence of organic solvent, the $\Delta H_{tot}^{o}(dried)$ and $\Delta H_{tot}^{o}(hydrated)$ values should be compared with water sorption data obtained independently, i.e., m_w^{tr} (Eq. 11). The dehydration enthalpy, ΔH_{deh}, includes the contributions resulting from (1) the difference between the partial enthalpies of the initially dried and hydrated enzyme and (2) the difference between the partial enthalpies of water in the enzyme and liquid phase.

$$\Delta H_{deh} = \frac{\Delta H^o_{tot}(hydrated) - \Delta H^o_{tot}(dried)}{m_w^{tr}} =$$

$$\frac{m_E}{m_w^{tr}}[\overline{H}_E^o - \overline{H}_E^h] + [\overline{H}_{w,\ liquid} - \overline{H}_{w,\ gas}] + \tag{11}$$

$$[\overline{H}_{w,\ gas} - \overline{H}_{w,\ \cdot initial.enzyme}] =$$

$$\Delta H^{H_2O/gas}_{desorp} + \Delta H^{H_2O/S}_{solv}$$

where $\overline{\boldsymbol{H}}_E^o$ is the partial enthalpy of the initially dried enzyme, $\overline{\boldsymbol{H}}_E^h$ is the partial enthalpy of the initially hydrated enzyme, and $\Delta \boldsymbol{H}^{H_2O/gas}_{desorp}$ is the molar enthalpy of transfer of water from the enzyme phase to the gas phase.

Eq. 11 is based on the assumption that the transfer of water from the enzyme phase to anhydrous organic solvent is complete.

4. Results and Discussion

4.1. Interaction Enthalpies of the Dried and Hydrated Enzyme with Organic Solvents

Figure 1 shows how the ΔH^0_{tot} values for the dried and hydrated enzyme preparations depend on the initial water content. The interaction enthalpies of the dried and hydrated CT with pure water are also given in Figure 1.

The interaction enthalpies depend in a complicated way on the initial water content. The interaction enthalpies (ΔH^0_{tot}) of the dried chymotrypsin with organic solvents are exothermic (negative). Immersion of the dried chymotrypsin in anhydrous organic liquids under study results in significant structural changes [15]. For example, alcohols result not only in generating some helical structure but also in the subsequent formation of more extensive β-sheets. Interaction of the dried CT with DMSO shows, as initial step, relatively fast randomizing of the enzyme structure followed by slow conversion to extensive β-aggregation.

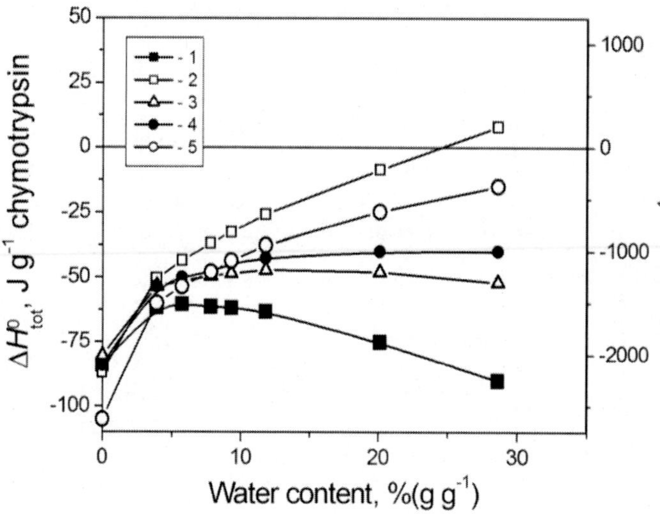

Figure 1. **Interaction enthalpies of the initially dried and hydrated** α-chymotrypsin with water and anhydrous organic solvents (ΔH^0_{tot}) as a function of water content: (1) DMSO, (2) formamide, (3) methanol, (4) ethylene glycol; (5) water.

At low water content (h_{ini} = 0-0.1 g water/g enzyme), there is a sharp change in the interaction enthalpies. It is well known that at a hydration level between 0 and 0.1 g water/g enzyme, water is strongly bound to ionizable and polar groups [1,2]. This implies that the primary dehydration process (the dehydration of polar and ionizable protein groups) is sufficient for the expression of the interaction enthalpies in this water content range. At high water content, the water-organic solvent interactions play a dominant role in the interaction enthalpies. For example, at high water content (h_{ini}>0.2 g water/g enzyme), the interaction enthalpies for formamide are endothermic (positive) (Figure 1). This result is due to the endothermic enthalpies of solution of water in this organic solvent. The enthalpy of solution of water in formamide at infinite dilution was found to be 1.0 kJ/mol water [23]. The interaction enthalpies for methanol, ethylene glycol, and DMSO are exothermic at high water content (h_{ini}>0.1 g water/g enzyme): (Figure 1). This result is due to the exothermic enthalpies of solution of water in these organic solvents [23,24].

4.2. Heat Effects of Dehydration of Chymotrypsin in Organic Solvents

The next step of this study was to calculate the heat effects of dehydration of chymotrypsin in organic solvents and determine the effect of the solvation ability of organic solvents on the heat effects of dehydration. We compared heat effects of two types: the interaction enthalpies for the dried ($\Delta H^0_{tot}(\text{dried})$) and hydrated enzymes ($\Delta H^0_{tot}(\text{hydrated})$) with anhydrous organic media. The molar enthalpies of dehydration of enzymes in organic solvents, ΔH_{deh}, were calculated by a modified version of Eq. 11, as given in Eq. 12:

$$\Delta H_{deh} = \frac{(\Delta H^0_{tot}(\text{hydrated}) - \Delta H^0_{tot}(\text{dried})) * 18 * 100}{h_{ini}} \quad (12)$$

Table 1. Heat effects of dehydration α-chymotrypsin in organic solvents, ΔH_{deh}, at 25°C (kJ/mol water)

Solvent	h_{ini}= 3.9%	h_{ini}= 5.7%	h_{ini}= 7.8%	h_{ini}= 9.7%	h_{ini}= 11.8%	h_{ini}= 20.1%	h_{ini}= 28.6%
Formamide	16.8 (0.6)	13.7 (0.5)	11.5 (0.4)	10.5 (0.4)	9.3 (0.3)	7.0 (0.2)	6.0 (0.2)
Ethylene Glycol	14.1 (0.5)	10.7 (0.4)	8.3 (0.2)	7.6 (0.2)	6.3 (0.2)	3.9 (0.2)	2.8 (0.2)
Methanol	12.4 (0.5)	9.1 (0.3)	7.1 (0.2)	6.2 (0.2)	5.1 (0.2)	2.9 (0.1)	1.8 (0.1)
DMSO	10.4 (0.3)	7.5 (0.2)	5.2 (0.2)	4.3 (0.2)	3.2 (0.2)	0.8 (0.1)	-0.7 (0.1)

where h_{ini} is the initial water content of chymotrypsin or lysozyme (%, g water/g enzyme), and the number 18 is the molecular weight of water (g/mol). The ΔH_{deh} values are summarized in Table 1.

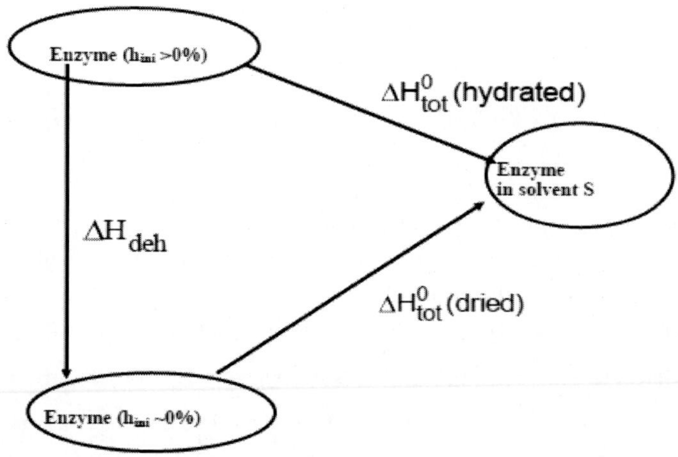

Figure 2. Scheme d escribing the interrelation between the interaction enthalpies of the dried, $\Delta H^0_{tot}(\text{dried})$, and hydrated, $\Delta H^0_{tot}(\text{hydrated})$, enzyme with anhydrous organic solvents and heat effects of dehydration, ΔH_{deh}.

A thermochemical model was proposed to explain the effect of organic solvents on the ΔH_{deh} values. This model is based on a thermochemical scheme which suggests that the interaction enthalpies of the hydrated enzyme, $\Delta H^0_{tot}(\text{hydrated})$, include the contributions from only two processes (Figure 2), i.e., (1) the transfer of water from the enzyme phase to the organic solvent phase, and (2) the interaction of the enzyme with organic solvent. Consequently, subtracting $\Delta H^0_{tot}(\text{dried})$ from $\Delta H^0_{tot}(\text{hydrated})$, one obtains the ΔH_{deh} values. Here, we assumed that the value of $\Delta H^0_{tot}(\text{dried})$ is the same for the dried and hydrated enzyme. It means that both the hydrated and dried enzyme samples must eventually pass into a state with the same enthalpy. In turn, the ΔH_{deh} value consists of the contributions from two processes (Figure 3), i.e., (1) the transfer of water from the enzyme phase to the gas phase, $\Delta H_{desorp}^{H_2O/gas.}$, and (2) the solvation of water by the organic solvent, $\Delta H_{solv}^{H_2O/S}$. Hence, the ΔH_{deh} value must depend linearly on the enthalpy of solvation of water, $\Delta H_{solv}^{H_2O/S}$ with the slope being close to unity, which is indeed the case, as shown in Figure 4. The parameters of the linear correlations, as outlined in Eq. 13, are given in Table 2.

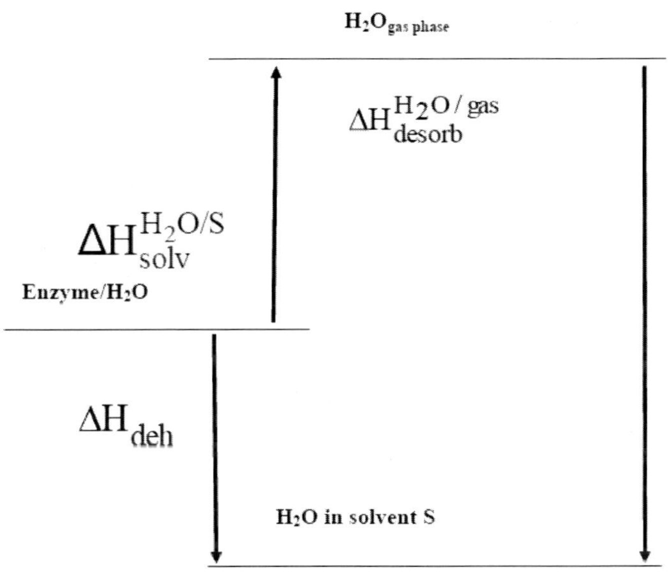

Figure 3. Scheme describing the separation of the heat effect of dehydration, ΔH_{deh}, of the hydrated enzyme (**Enzyme/H_2O**) by organic solvents into two contributions: (1) the transfer of water from the initially hydrated enzyme to the gas phase, $\Delta H_{desorb}^{H_2O/gas}$, and (2) the solvation of water by the solvent, $\Delta H_{solv}^{H_2O/S}$.

$$\Delta H_{deh} = \Delta H_{desorp}^{H_2O/gas} + \text{slope} * \Delta H_{solv}^{H_2O/S} \tag{13}$$

The intercept of Eq. 13 corresponds to the molar enthalpy of transfer of water from the enzyme phase to the gas phase $\Delta H_{desorp}^{H_2O/gas.}$. At hydration level ($h_{ini}$<0.1 g water/g enzyme), the strong hydration sites are preferentially occupied. These water binding sites are chemically diverse and include a variety of ionizable and polar functional protein groups. The most significant $\Delta H_{desorp}^{H_2O/gas.}$ values were observed in this range (Figure 5).

Table 2. Parameters of Eq. 13

h_{ini}, % (g water/g enzyme)	$\Delta H_{desorb}^{H_2O/gas}$ kJ/ mol water	Slope	R	N	Standard error of estimation, S_o
3.9 (0.1)	59.8 (2.1)	1.0 (0.1)	0.99	5	0.23
5.7 (0.2)	57.5 (2.0)	1.0 (0.1)	0.99	5	0.22
7.8 (0.2)	55.4 (1.6)	1.0 (0.1)	0.99	5	0.20
9.7 (0.2)	53.8 (1.7)	1.0 (0.1)	0.99	5	0.19
11.8 (0.2)	52.4 (1.6)	1.0 (0.1)	0.99	5	0.18
20.1 (0.3)	49.6 (1.4)	1.0 (0.1)	0.99	5	0.16
28.6 (0.4)	48.3 (1.3)	1.0 (0.1)	0.99	5	0.14

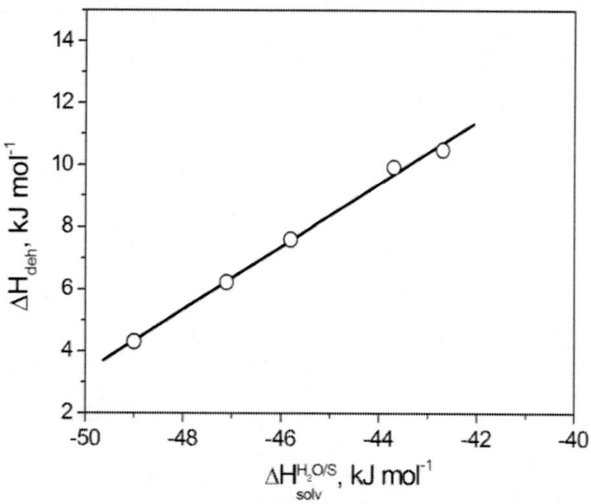

Figure 4. The enthalpy of dehydration of α-chymortypsin, ΔH_{deh}, (initial water content h_{ini}=9.7% (g/g)) as a function of the enthalpy of solvation of water, $\Delta H_{solv}^{H_2O/S}$, at infinite dilution and 25°C: (1) DMSO, (2) methanol, (3) ethylene glycol, (4) water, (5) formamide.

Table 3. List of symbols and abbreviations

Symbol	Name	Unit
ΔH^{0}_{tot} (dried)	Interaction enthalpy of the initially dried enzyme with anhydrous organic solvent	J
ΔH_{tot} (dried)	Interaction enthalpy of the initially dried enzyme with water-organic mixture	J
ΔH^{0}_{tot} (hydrated)	Interaction enthalpy of the initially hydrated enzyme with anhydrous organic solvent	J
ΔH_{tot} (hydrated)	Interaction enthalpy of the initially hydrated enzyme with water-organic mixture	J
$\Delta H^{H_2O/S}_{sol}$	Enthalpy of solution of water in a solvent at infinite dilution and 25°C	kJ/mol water
$\Delta H^{H_2O}_{vap}$	Enthalpy of vaporization of water at atmospheric pressure and 25°C	kJ/mol water
$\Delta H^{H_2O/S}_{solv}$	Enthalpy of solvation of water in organic solvent at infinite dilution and 25°C or enthalpy of solvation of water in pure liquid water at 25°C	kJ/mol water
h_{ini}	Initial water content of chymotsypsin or lysozyme	% (g water/g enzyme)
h_{res}	Residual water content bound to enzyme immersed in organic solvent	% (g water/g enzyme)
ΔH_{deh}	Enthalpy of dehydration of the initially hydrated enzyme in organic solvents (enthalpy of transfer water from the enzyme phase to the organic solvent liquid phase)	kJ/mol water
$\Delta H^{H_2O/gas.}_{desorp}$	Enthalpy of transfer of water from the initially hydrated enzyme to the gas phase	kJ/mol water

At high hydration level (h_{ini}>0.1 g water/g enzyme), the $\Delta H^{H_2O/gas.}_{desorp}$ values do not depend markedly on the water content. At h_{ini}>0.1 g/g, the

$\Delta H_{desorp}^{H_2O/gas.}$ values reach saturation and close to the enthalpy of vaporisation of water (43.7 kJ/mol water).

To test the reliability of our findings, we compared the $\Delta H_{desorp}^{H_2O/gas.}$ values determined in this study with previously published results [27,28] on the interaction enthalpies of the dried and hydrated chymotrypsin with pure water. The $\Delta H_{desorb}^{H_2O/gas}$ values were calculated by Eq. 14:

$$\Delta H_{desorp}^{H_2O/gas} = \frac{(\Delta H_{tot}^{0}(\text{hydrated}) - \Delta H_{tot}^{0}(\text{dried}))*18*100}{h_{ini}} + \Delta H_{vap}^{H_2O} \quad (14)$$

where $\Delta H_{vap}^{H_2O}$ is the enthalpy of vaporization of water at 25°C and atmospheric pressure (kJ/mol water), h_{ini} is the initial water content of chymotrypsin (%, g water/g protein), the number 18 is the molecular weight of water is the molecular weight of water (g/mol), $\Delta H_{tot}^{0}(\text{hydrated})$ is the interaction enthalpy of the hydrated enzyme with pure liquid water at 25°C (J g^{-1} enzyme). As shown in Figure 5, the molar enthalpies of transfer of water from the initially hydrated CT to the gas phase calculated using Eq. 13 agree well with those obtained using Eq. 14.

As concluded from Table 1, the ΔH_{deh} values may be endothermic or exothermic depending on the initial water content and the water solvation enthalpy value. For example, the ΔH_{deh} values for formamide, methanol, and ethylene glycol are endothermic. This result is due to the fact that the $\Delta H_{desorb}^{H_2O/gas}$ value is higher than the enthalpy of desolvation of water in these liquids in the entire range of water contents. In order to illustrate this situation, a line corresponding to the enthalpy of desolvation of water in formamide is given in Figure 5.

DMSO has the most exothermic water solvation enthalpy value from the solvents under study. The ΔH_{deh} values for DMSO are endothermic at low water contents. However, the ΔH_{deh} values for DMSO are exothermic at

h_{ini} =29.0% (g g^{-1}). In order to illustrate this situation, a line corresponding to the enthalpy of desolvation of water in DMSO is presented in Figure 5. This result is due to the fact that at the highest water contents, the $\Delta H^{H_2O/gas}_{desorb}$ value is lower than the enthalpy of desolvation of water in DMSO. These facts imply that the average water binding affinity in the hydrated proteins may be lower than in the good solvating liquids like DMSO that can form strong hydrogen bonds with various hydrogen bond donors.

CONCLUSION

A special thermochemical scheme was proposed to predict the effect of the solvation ability of a medium on the molar enthalpy of transfer of water from the enzyme phase to the organic solvent liquid phase. This scheme shows that the molar enthalpy of enzyme dehydration in organic solvents may be predicted on the basis of the enthalpies of solution of the hydrated ($\Delta H^0_{tot}(\text{hydrated})$) and dried ($\Delta H^0_{tot}(\text{dried})$) protein preparations in pure water and the enthalpies of solvation of water in the corresponding solvents at infinite dilution, $\Delta H^{H_2O/S}_{solv}$. The results obtained give strong support to the ideas that (1) the changes in the dehydration enthalpies are mainly determined by the solvation of water in hydrophilic solvents and (2) the data of the enthalpies of solvation of water by the solvent, $\Delta H^{H_2O/S}_{solv}$, at infinite dilution and room temperature reflect this effect.

REFERENCES

[1] Klibanov, A.M. (2001) Improving enzymes by using them in organic solvents, *Nature,* 409, 241-246.

[2] Carrea, G., Riva, S. (2000) Properties and synthetic applications of enzymes in organic solvents, *Angew. Chem. Int. Ed.,* 39, 2226-2254.

[3] Gupta, M.N. (2000) *Methods in Non-Aqueous Enzymology,* (M.N. Gupta, ed.), Birhäuser Verlag, Basel-Boston-Berlin.

[4] Ke, T., Klibanov, A.M. (1998) On enzymatic activity in organic solvents as a function of enzyme history, *Biotech. Bioeng. 57, 746-750.*

[5] Halling, P.J. (2004) What can we learn by studying enzymes in nonaqueous media? *Philos. Trans. R. Soc. Lond. B Biol. Sci.,* 359, 1287– 1297.

[6] Micaelo, N.M., Soares, C.M. (2007) Modeling hydration mechanisms of enzymes in nonpolar and polar organic solvents, *FEBS Journal*, 274, 2424–2436

[7] Sirotkin, V.A. (2005) Effect of dioxane on the structure and hydration-dehydration of α-chymotrypsin as measured by FTIR spectroscopy. *Biochim. Biophys. Acta*., 1750, 17-29.

[8] Sirotkin V.A., Solomonov B.N., Faizullin D.A., Fedotov V.D.(2002) Sorption of water and acetonitrile by human serum albumin. *Russ. J. Phys. Chem. A,* 76, 2051-2057.

[9] Smith A.L., Shirazi H.M., Mulligan S.R. (2002) Water sorption isotherms and enthalpies by lysozyme using the quartz crystal microbalance/heat conduction calorimeter. *Biochim. Biophys. Acta,* 1594, 150-159.

[10] Kocherbitov V., Arnebrant T., Söderman O. (2004) Lysozyme-water interactions studied by sorption calorimetry. *J. Phys. Chem. B,* 108, 19036-19042.

[11] Borisover, M.D., Sirotkin, V.A., Solomonov, B.N. (1995) Thermodynamics of water binding by human serum albumin suspended in acetonitrile. *Thermochim. Acta*, 254, 47-53.

[12] Sirotkin V.A., Korolev D.V. (2005) Effect of acetonitrile on the hydration of human serum albumin films: a calorimetric and spectroscopic study. *Thermochim. Acta,* 432, 246-253.

[13] Sirotkin, V.A., Faizullin, D.A. (2004) Interaction enthalpies of solid human serum albumin with water-dioxane mixtures: comparison with water and organic solvent vapor sorption, *Thermochim. Acta*, 415, 127-133.

[14] Gorman, L.S., Dordick, J.S. (1992) Organic solvents strip water off enzymes. *Biotechnol. Bioeng.,* 39, 392-397.

[15] Sirotkin, V.A., Zinatullin, A.N. Solomonov,B.N., Faizullin, D.A., Fedotov, V.D. (2002) Interaction enthalpies of solid bovine pancreatic chymotrypsin with organic solvents: comparison with FTIR-spectroscopic data. *Thermochim. Acta*, 382, 151-160.

[16] Fersht A. Structure and mechanism in protein science: a guide to enzyme catalysis and protein folding. New York: Freeman & Co, 1999.

[17] Lehninger AL, Nelson DL, Cox MM. Principles of Biochemistry. New York: Worth, 1993.

[18] Perrin, D.D., Armarego, W.L.F., Perrin, D.R. (1980) *Purification of Laboratory Chemicals*, Oxford: Pergamon Press.

[19] Borisover, M.D., Sirotkin, V.A., Zakharychev, D.V., Solomonov, B.N. (2001) Calorimetric methods in evaluating hydration and solvation of solid proteins immersed in organic solvents, In: *Enzymes in Nonaqueous Solvents* (Vulfson, E.N., Halling, P.J., editors), Humana Press, Totowa, 183-202.

[20] Nikol'skii, B. P. (1963). *Chemist's Handbook,* Goskhimizdat: Leningrad, Russia.

[21] Greenspan, L. (1977). Humidity fixed points of binary saturated aqueous solutions. *J Res. Natl. Bur. Stand., Sect. A*, 81**,** 89-96.

[22] Kogan, V.B., Fridman, B.N., Kafarov, V.V. (1966) *Equilibrium between Liquid and Vapor. Handbook*, Nauka, Moscow.

[23] Belousov, V.P., Panov, M.Y. (1994) *Thermodynamic properties of aqueous solutions of organic substances*. Boca Raton, Fla.: CRC Press.

[24] Belousov, V.P., Morachevski, A.G. (1970) *Heats of Mixing of Liquids*, Khimiya, Leningrad.

[25] Gregory, R. B. (1995). Protein hydration and glass transition behavior. In *Protein-Solvent Interactions* (Gregory, R. B., Ed.), Marcel Dekker: New York pp. 191-264.

[26] Rupley, J.A., Careri, G. (1991) Protein hydration and function. *Adv. Protein Chem,* 41, 37-172.

[27] Sirotkin, V.A., Khadiullina, A.V. (2011) Hydration of proteins: excess partial enthalpies of water and proteins. *J. Phys. Chem. B*. 115**,** 15110-15118.

[28] Sirotkin, V.A.; Khadiullina, A.V. (2011) Hydration of chymotrypsin: excess partial enthalpies of water and enzyme. *Thermochim. Acta*. 522**,** 205-210.

In: Protein – Water Interactions
Editor: Vladimir A. Sirotkin
ISBN: 978-1-63463-007-8
© 2014 Nova Science Publishers, Inc.

Chapter 7

ANALYSIS OF THE ORGANIC SOLVENT EFFECT ON THE HYDRATION AND STRUCTURE OF ALPHA-CHYMOTRYPSIN BY FTIR SPECTROSCOPY

Vladimir A. Sirotkin*
Kazan Federal University, A.M. Butlerov Institute of Chemistry,
Kazan, Russia

ABSTRACT

The effect of acetonitrile (AN) on the hydration and structure of bovine pancreatic α-chymotrypsin (CT) was characterized using FTIR spectroscopy at 25°C. The obtained results show that the hydration and structure of α-chymotrypsin depend markedly on how enzyme has been hydrated – whether in the presence or in the absence of organic solvent.

Acetonitrile decreases markedly the water content at a given water activity (a_w). This behavior suggests that the suppression in the uptake of water is due to a competition for water-binding sites on α-chymotrypsin by acetonitrile. Changes in the structure of CT were determined from infrared spectra by analyzing the structure of amide I band. It was found that, at low water activity, the protein-protein contacts in the dried enzyme largely govern its thermodynamic, structural, and functional properties. At high water activity, the chymotrypsin state is determined

* E-mail: vsir@mail.ru.

by the enzyme-organic solvent and enzyme-water interactions resulting in the enzyme denaturation. The interplay of these two factors is responsible for the observed complex shape of the enzyme storage stability - a_w curves, with an extremum at intermediate water activity values.

Keywords: Enzyme hydration and structure, Bovine pancreatic α-chymotrypsin, Biocatalysis in organic solvents, Acetonitrile, Fourier transform infrared spectroscopy (FTIR spectroscopy)

1. INTRODUCTION

Infrared spectroscopy is one of the effective methods for analyzing the structure of proteins in various environments, including aqueous and non-aqueous media [1-9]. This method has been successfully used in studying the secondary structure of proteins in various states, including protein films with various degrees of humidity and solid protein preparations immersed into pure organic solvents and water - organic mixtures. Infrared spectroscopy is also effective in studying the hydration of proteins.

By means of the IR spectroscopic measurements the structure and stability of the dehydrated α-chymotrypsin were examined in a series of anhydrous organic liquids, including hydrocarbons, alcohols and hydrogen bond accepting solvents [6]. It was shown that solvent potential to form hydrogen bonds appears to be an important factor controlling the stability of dehydrated proteins in organic media. The effect of organic solvents on the hydration and structure of proteins was recently studied by isothermal calorimetry and IR spectroscopy [7-9].

In this study the effect of organic solvent on the enzyme hydration was investigated by FTIR spectroscopy in the water activity range from 0 to 1.0 at 25°C. The sorption data were compared with the corresponding structural changes that occur on the interaction of the enzyme with water and organic molecules. The aim of this FTIR spectroscopic study is to elucidate the effect of organic solvent molecules on the hydration, structure, and functional properties of the enzyme.

Bovine pancreatic α-chymotrypsin was used as a model enzyme. It is one the most studied and applied in biochemistry, molecular biology and enzymology [10,11]. Physiological role of CT is to hydrolyze peptide bonds [10,11]. α-Chymotrypsin is an example of a predominantly β-sheet protein.

The choice of acetonitrile was determined by the following reasons:

A) Acetonitrile is a water-miscible organic solvent. Therefore, it is possible to study the effects of this low molecular organic substance on the hydration and structure of α-chymotrypsin over the whole range of thermodynamic water activity.

B) It is capable of forming strong hydrogen bonds with various hydrogen donors. However, in contrast to water, it has no evident hydrogen bond donating ability.

2. EXPERIMENTAL

2.1. Materials

Bovine pancreatic α-chymotrypsin (Sigma, No. C 4129, essentially salt free; EC 3.4.21.1; specific activity of 66 units/mg of solid) was used without further purification. Acetonitrile (reagent grade, purity >99%) was purified and dried according to the recommendations [12]. Water used was doubly distilled.

2.2. Thermodynamic Activity of Water

Water activity (a_W) in organic solvent was calculated using the Eq. 1:

$$a_w = \gamma_w x_w \tag{1}$$

where x_W is the mole fraction of water in the solution and γ_W is the activity coefficient of water (in mole fractions; the standard state is pure water).

Water activity coefficients (γ_W) were calculated using literature data [13] on the vapor-liquid equilibrium according to Eq. 2.

$$\gamma_w = \frac{y_w P_{tot}}{x_w P_w^o} \tag{2}$$

where y_W is the measured mole fraction of water in vapor phase, P_t is the total pressure, P_o is the saturated vapor pressure of pure water at the same temperature and x_W is the mole fraction of water in the liquid phase.

Figure 1 shows the thermodynamic activity of water as a function of the water mole fraction in organic solvent at 25°C.

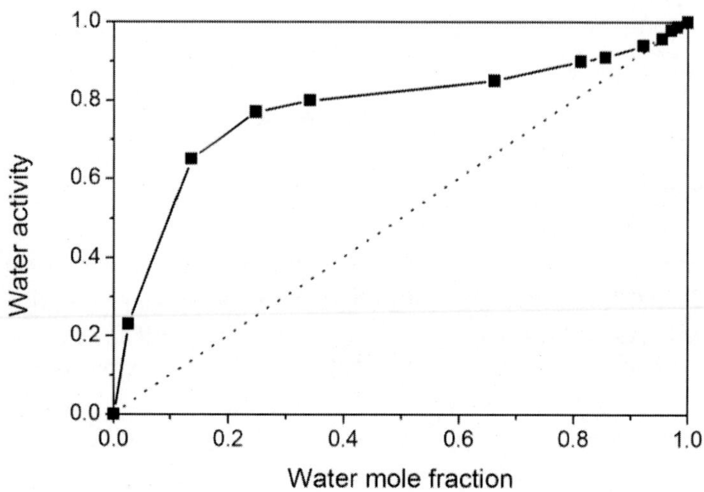

Figure 1. Thermodynamic activity of water plotted versus the water mole fraction in organic solvent at 25°C. The dashed lines correspond to the ideal binary mixture.

2.3. Spectroscopic Measurements

The FTIR spectra were recorded with a MAGNA 550 spectrometer from Nicolet. The infrared spectra were measured at 25°C as described previously [6,7]. Each spectrum was obtained by coadding 256 scans at a spectral resolution of 2 cm^{-1}. The sample chamber was purged with dry and carbon dioxide free air.

The infrared spectra were obtained with glassy like protein films casted from 2% (w/v) water solution onto the CaF_2 window at room humidity and temperature. Schematic representation of the experimental set-up for FTIR spectroscopic measurements of the structural and sorption characteristics of proteins is given in [9]. After mounting the window in the sample cell, the film was dehydrated by flushing air dried over P_2O_5 powder. Relative water vapour pressure over P_2O_5 at 25°C does not exceed 0.01 [14]. The protein films were flushed until no further spectral changes were detected in the 3450 cm^{-1} water absorbance region and amide A contour on this side represented a smooth line without any visible shoulders. The spectrum of this sample was used as a reference spectrum for calculation of the difference spectra.

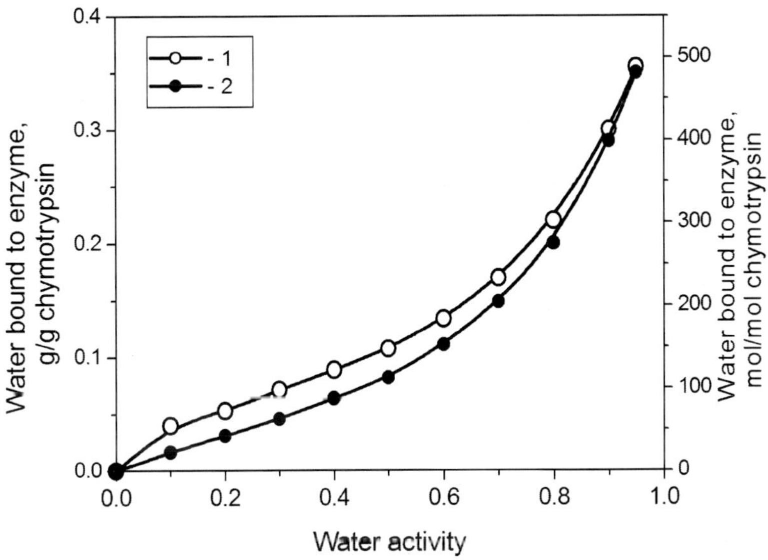

Figure 2. (1) Water sorption isotherm for α-chymotrypsin in the absence of acetonitrile; (2) Water sorption isotherm for α-chymotrypsin in the presence of acetonitrile. Solid lines were fitted by a set of polynomials.

The initially dehydrated enzyme was exposed to pure water vapor or water-organic vapor mixtures. In the first case, pure water vapor consecutively flowed through the thermostated glass tube with drying agent (P_2O_5), saturator filled with pure liquid water and then through the measuring cell containing the protein sample. The temperature of sample cell was 25°C. The water activity (a_w) in vapor phase was adjusted by changing the difference between the temperature of the saturator and cell. Data on water vapor pressure at various temperatures were taken from [14].

In the second case, the air consecutively flowed through the thermostated glass tube with drying agent (P_2O_5), saturator filled with a water – organic mixture and then through the sample cell. The temperature of the sample cell and saturator were 25°C. The water activity (a_w) in vapor phase was adjusted by changing the water activity in the liquid water – organic mixture.

Water sorption by protein films was controlled in the region of OH stretching vibration band at 3450 cm^{-1}, which is the most intensive band of sorbed water.

Enzyme films were flushed by pure water vapour or water – organic vapour mixtures until no further changes were detected in the absorbance values at 3450 cm^{-1} (water sorption control). Upon hydration in the absence of organic solvent no noticeable spectral changes were observed after 60 min. Upon hydration in the presence of organic solvent, the sorption equilibrium was reached after 180 min. This increase in the time required for the attainment of sorption equilibrium in the presence of organic solvent is presumably due to pronounced conformational rearrangements and considerable changes in the mechanism of water sorption.

Supposing that during sorption the enzyme film swells predominantly due to increasing thickness, the enzyme hydration was calculated from Eq. 3:

$$h = \frac{2.303 \cdot S_{water} \varepsilon_{CT}}{B_{water} D_{CT}} \qquad (3)$$

where h is the enzyme hydration, mol water mol^{-1} enzyme, S_{water} is the area of water absorbance band, cm^{-1}; ε_P is the enzyme molar extinction at the maximum of Amide I band, l mol^{-1} protein cm^{-1}; B_{water} is the water integral molar extinction coefficient, l mol^{-1} water cm^{-2}; D_{CT} is the optical density at the maximum of amide I band. For pure water, it was taken that B_{water} = 96,000±1000 l mol^{-1} water cm^{-2} [15].

The molar absorption extinction coefficient of α-chymotrypsin was determined measuring the amide I spectra of protein solutions in heavy water. The molar absorption extinction coefficient of chymotrypsin (ε_{CT}) is 80,000±200 L mol^{-1} cm^{-1}. The molecular weight of bovine pancreatic α-chymotrypsin was taken as 25,000 Da. The reliability of water sorption measurements by FTIR spectroscopy was shown in [9].

3. Results and Discussion

3.1. Water sorption Isotherms

Figure 2 shows the water sorption isotherms for α-chymotrypsin in the presence and absence of AN. The initial state of the protein film (**zero**

hydration level) was obtained by drying in air at water activity less than 0.01 (Film A). On maintaining this "dry" sample in equilibrium with a vacuum of 0.1 Pa and at 298 K for three hours, it lost about 0.3 % of its weight, which for α-chymotrypsin implies that at the zero hydration level there are about three water molecules strongly bound to each protein molecule. As shown in Figure 3, the water sorption isotherms resemble sigmoid shaped curves. This result is in accordance with those for α-chymotrypsin and other proteins [1,2].

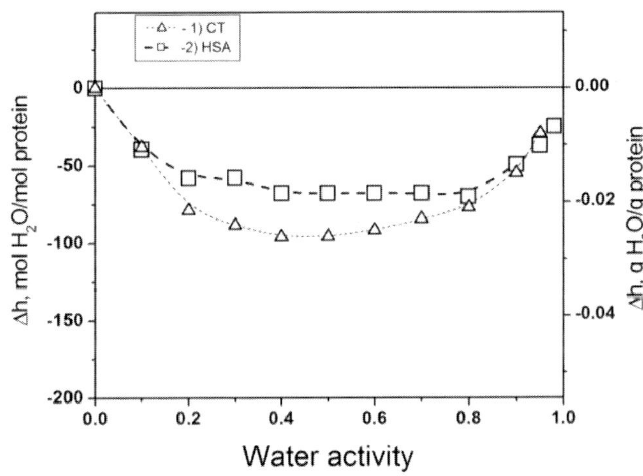

Figure 3. Differences in the water uptake between the water sorption isotherms in the presence and absence of acetonitrile.

The presence of organic molecules markedly affects the ability of CT to bind water. The effect of acetonitrile on the water sorption was characterised by the difference, Δh, in water uptake in the presence and absence of organic solvent (Figure 3). As shown in Figure 3, a considerable decrease in the uptake of water was observed over the whole range of water activity. This behavior suggests that the suppression in the water uptake is due to a competition for water binding sites on CT by acetonitrile molecules. A sharp decrease in the Δh values was found at $a_w>0.8$. This decrease agrees well with the cooperative increase of the fraction of the H-bond associated water molecules in the presence of acetonitrile [15].

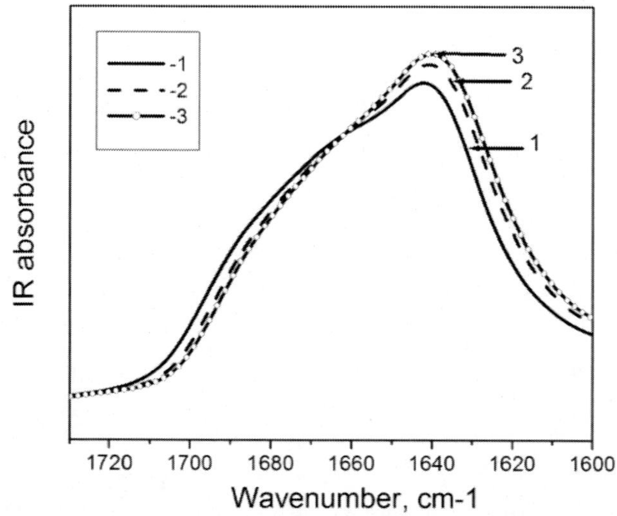

Figure 4. Typical absorbance spectra of chymotrypsin in the amide I region in the absence of AN: (1) a_w= 0.01; (2) a_w=0.2; (3) a_w=0.8.

To confirm the generality of our finding, the Δh values for another globular protein, human serum albumin, were also presented in Figure 3. As concluded from Figure 3, the effect of acetonitrile on water binding by HSA is similar to that observed for α-chymotrypsin.

The BET model [16] is widely used for describing the sorption ability of various solids, including proteins [1,2,6,9, 16,17] (Eq. 4):

$$h = h_m \left[\frac{K a_w}{1 + K a_w} + \frac{a_w}{1 - a_w} \right] \tag{4}$$

where h is the hydration of solid enzyme (mol water per mol enzyme); h_m is the number of water binding sites (mol water per mol enzyme); K is the equilibrium water sorption constant.

It was found that the BET equation (Eq. 4) describes the isotherms displayed in Figure 2 up to a water activity no more than 0.6. This result is in close agreement with the data obtained in previous studies on α-chymotrypsin and other proteins [1,2,6,9, 16,17].

The water sorption parameters estimated from Eq. 4 are presented in Table 1. As shown in Table 1, the value of sorption constant K is markedly smaller in the presence of acetonitrile. This implies that acetonitrile molecules suppress markedly the water sorption on the CT films. Similar results were observed for human serum albumin [7].

Table 1. Parameters of water sorption by chymotrypsin and human serum albumin estimated by Eq. 4

*h*m, mol water /mol protein	K	s_o^2
In the absence of acetonitrile		
Chymotrypsin [this study]		
79.3 (1.2)	18.1 (3.0)	1.9
Human serum albumin [7]		
174.3 (3.1)	12.1 (4.3)	3.3
In the presence of acetonitrile		
Chymotrypsin [this study]		
68.5 (1.5)	3.3 (0.4)	2.6
Human serum albumin [7]		
149.7 (4.0)	4.0 (0.8)	7.1

Notes:

1) The water activity range of applicability is 0 - 0.6.

2) s_o is the residual standard deviation.

3) The values of confidence interval of the parameters calculated by Eq. 4 are given in parentheses.

3.2. Analysis and Band Assignment of Protein Infrared Spectra

Infrared spectra of proteins contain some vibrational bands arising from the amide groups of protein backbone. Among them, the amide I band between 1700 and 1600 cm^{-1}, which is due primarily to the C=O stretch of the peptide linkages, is the most sensitive to changes in the secondary structure of proteins. The infrared spectra of α-chymotrypsin in the amide I region are presented in Figures 4 and 5. Figures 4 and 5 show typical absorbance and second derivative spectra of α-chymotrypsin in the absence of acetonitrile in the amide I region. The amide I contour consists of a number of bands at frequencies that are characteristic of specific types of secondary structure. It has been well documented that the frequency at maximum absorbance of the amide I band is determined by the predominant form of secondary structure in the protein [3-5]. The maximum absorbance of the amide I band of proteins with a high content of α-helix occurs between 1658 and 1654 cm^{-1} and with a high content of β-sheet structure the maximum is found between 1640 and 1635 cm^{-1}.

As shown in Figures 4 and 5, the most dominant band component of the chymotrypsin spectra is the band at 1639 cm^{-1}, which is attributed to the native β-sheet structure [3-5]. The band at 1685 cm^{-1} was assigned to β-sheet structure [3-5]. Minor component in the 1670-1665 cm^{-1} region was assigned to irregular secondary structures (β-turns and extended chains) [3-5]. The band at 1700 cm^{-1} was assigned to the non-ionized H-bonded COOH groups [19]. These groups are strongly intramolecularly hydrogen bonded. Upon hydration (Figure 5) these groups forms more weak H-bonds with water molecules (broad band at 1712 cm^{-1}).

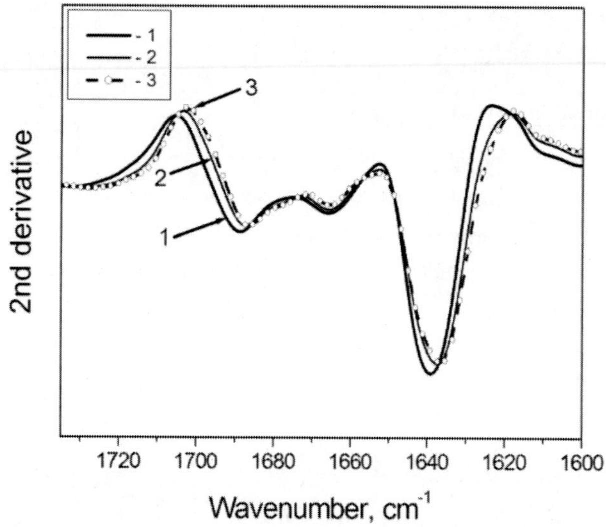

Figure 5. Second-derivative spectra of chymotrypsin in the amide I region in the absence of AN: (1) a_w= 0.01; (2) a_w=0.2; (3) a_w=0.8. All the spectra were normalised to the amide I band intensity of a Film A at water activity of 0.01.

Figures 6 and 7 show the absorbance and second derivative spectra of α-chymotrypsin in the presence of acetonitrile in the amide I region. As concluded from Figures 6 and 7, the spectra of chymotrypsin are very different from those in the absence of acetonitrile.

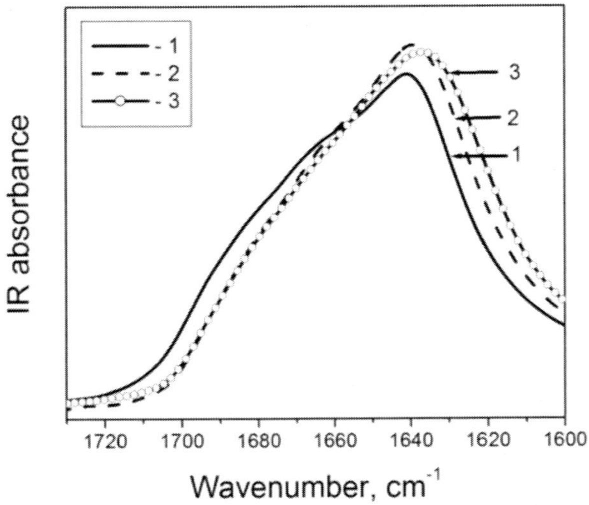

Figure 6. Typical absorbance spectra of chymotrypsin in the amide I region in the presence of AN: (1) a_w= 0.01; (2) a_w=0.5; (3) a_w=0.8.

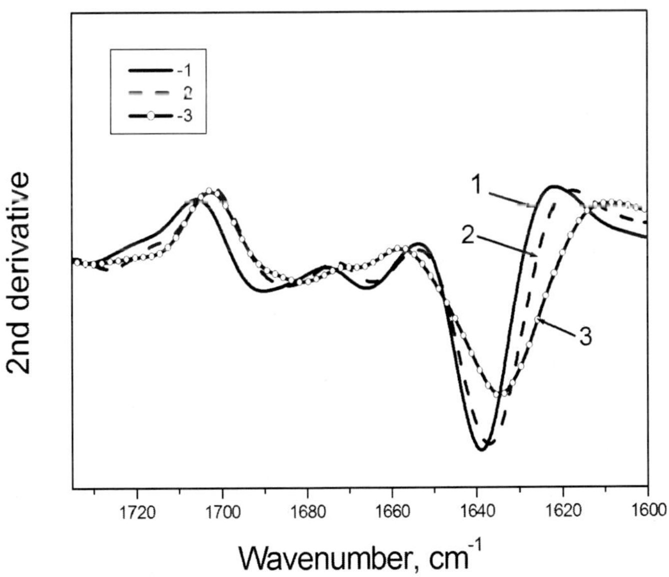

Figure 7. Second-derivative spectra of chymotrypsin in the amide I region in the presence of AN: (1) a_w= 0.01; (2) a_w=0.5; (3) a_w=0.8. All the spectra were normalised to the amide I band intensity of a Film A at water activity of 0.01.

As the water activity was raised from 0.01 to 0.98, the amide I absorbance maxima was red-shifted about 6 cm^{-1} to near 1633 cm^{-1}. This broad band at 1633 cm^{-1} (Figures 6 and 7, spectrum 2) may be caused by the mixing of the native β-sheet band (1639 cm^{-1}) with an intermolecular β-sheet band (1628 cm^{-1}) [3-5, 9].

As concluded from Figure 7, upon hydration in the presence of AN, the band at 1700 cm^{-1} disappears giving rise to two bands at 1712 cm^{-1} and 1724 cm^{-1}. The band at 1712 cm^{-1} was assigned to the non-ionized COOH groups forming H-bonds with water molecules. The band at 1724 cm^{-1} was assigned to the non-ionized COOH groups solvated by the organic solvent molecules. The band at 1724 cm^{-1} was not observed upon hydration in the absence of AN. The band at 1724 cm^{-1} disappears at the highest water activities.

The effect of acetonitrile on the CT structure was characterised by the difference functions. The D_x / D_o functions obtained in the presence of AN (Figure 8) have been compared with those determined for CT in the absence of AN. This comparison can show any effects of the organic solvent molecules on the enzyme structure. The differences between the D_x / D_o values obtained in the presence and absence of AN are presented in Figure 9.

As shown in Figure 9, the function is consistent with the changes in the values (Figure 3). This result indicates that the effect of AN on the enzyme structure is due to a competition for strong binding (polar/ionizable) sites on CT by organic solvent molecules. At the lowest water activity values, the values are close to zero (Figure 9, curve A). Hence, it may be concluded, that the presence of AN has little effect on the state of the intramolecularly H-bonded COOH groups in the initially dried enzyme.

The most pronounced effect of AN was observed for the hydrated CT in the water activity range of 0.6-0.8 (Figure 9). It is well known that at a hydration level between zero and 0.1 g/g, water is strongly bound to polar groups on the proteins [1,2]. This implies that the primary hydration process (the hydration of polar and ionizable protein groups) is sufficient for the expression of the acetonitrile-induced structural changes.

A sharp decrease in the values was found at a_w>0.8. This decrease agrees with the cooperative increase in the fraction of the H-bond associated water molecules [15] and increase in the values (Figure 3).

The band at 1626 cm^{-1} was assigned to intermolecular β-aggregates [9]. Upon hydration, in the water activity range from 0 to 0.5 (h<0.1 g/g), the $\Delta D_{1638} / D_o$, and values are close to zero (Figure 9). Minor decrease in the enzyme storage stability was observed at low water activity (Figure 10).

Hence, it may be concluded, that the presence of acetonitrile has little effect on the structure of the initially dried enzyme.

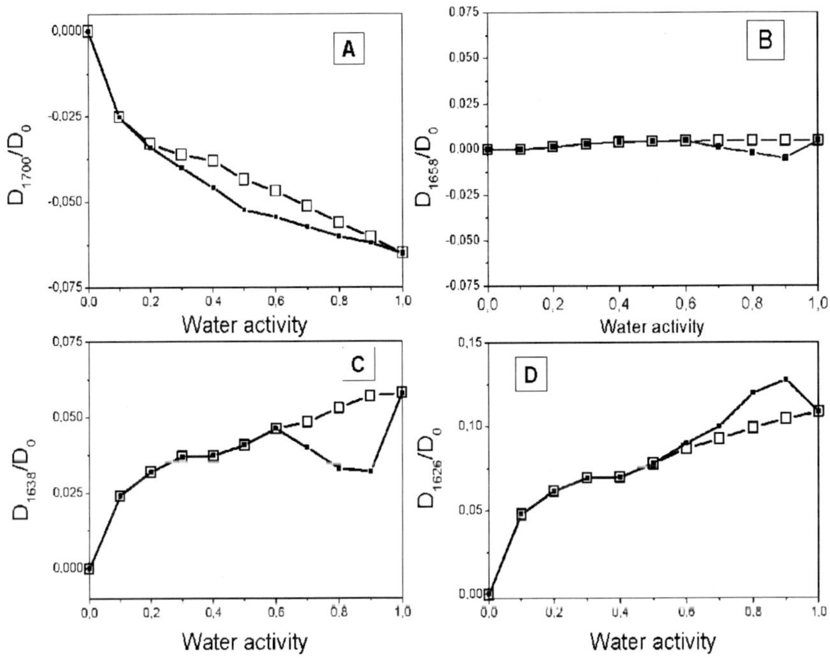

Figure 8. Absorbance at 1700 cm^{-1} (A), 1658 cm-1 (B), 1638 cm^{-1} (C), 1626 cm^{-1} (D) as a function of water activity: open squares - in the absence of acetonitrile and solid squares - in the presence of acetonitrile. D_o is the absorbance value for Film A at water activity of 0.01.

Above the threshold water activity of 0.5 (h>0.1 g/g), the values are sharply increased and reach a maximum at water activity of 0.8-0.9. This increase in the values at a_w~0.8-0.9 agrees well with a decrease in the and values (native β-sheet and α-helix structures, correspondingly) (Figure 9). These results indicate that the presence of AN has significant effect on the secondary structure of the hydrated enzyme.

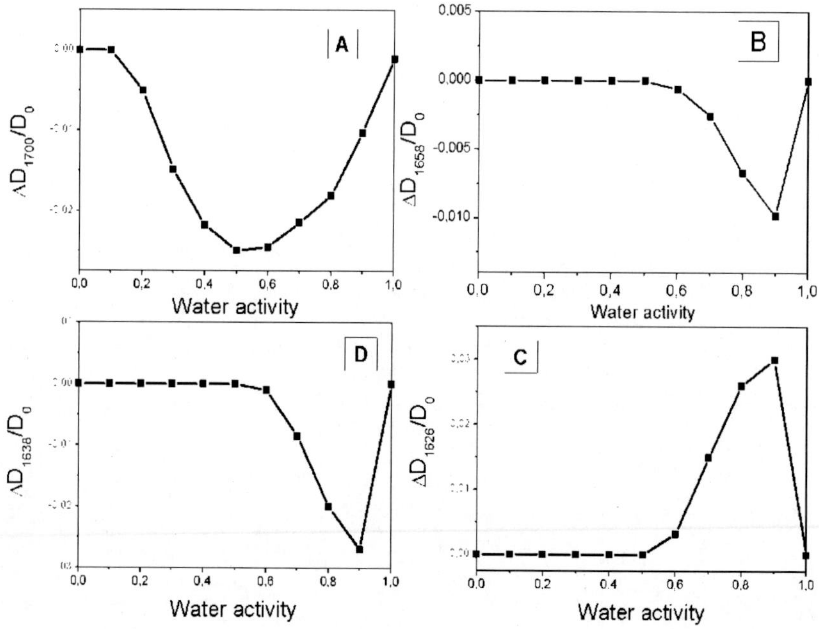

Figure 9. (A) Differences between the D_x/D_o values obtained in the presence and absence of acetonitrile as a function of water activity: 1700 cm^{-1} (A), 1658 cm-1 (B), 1638 cm^{-1} (C), 1626 cm^{-1} (D).

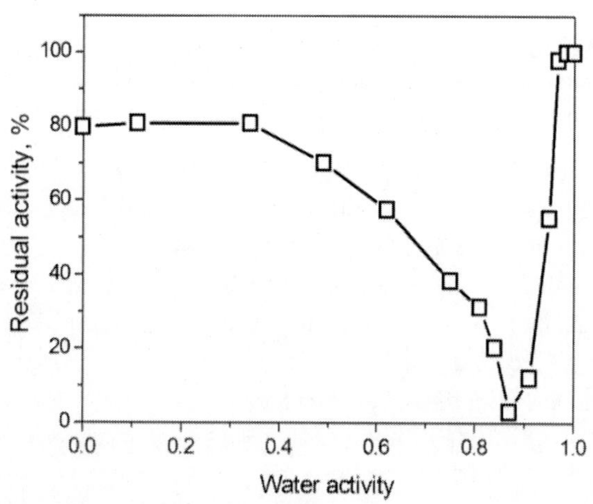

Figure 10. Residual activity of chymotrypsin in water-acetonitrile mixtures (Hydrolysis of N-acetyl-L-tyrosine ethyl ester by α-chymotrypsin preliminary incubated in water-acetonitrile mixtures). Adapted data from [20].

As concluded from Figure 9, the $\Delta D_{1638} / D_o$, and $\Delta D_{1658} / D_o$ functions are consistent with the changes in the chymotrypsin storage stability in water - acetonitrile mixtures (Figure 10). This means that these bands are sensitive to the effect of AN on the state of the active site conformation.

3.3. Effect of Acetonitrile on the Hydration and Structure of α-Chymotrypsin

3.3.1. Low Water Activity

Like for many proteins, dehydration of CT leads to the formation of protein-protein contacts due to proton-transfer phenomena and hydrogen bonding between polar and ionizable protein groups. These processes result in a rigid (glassy-like) structure in the dried state [1]. Therefore, certain moieties of the dried protein are unavailable for sorption due to strong interactions between them resulting in sorption hysteresis (for example, [9]).

It was previously shown that the potential of a solvent to form hydrogen bonds is an important factor that controls the state and structure of dehydrated proteins at room temperature [6]. Hence, H-donating and H-accepting properties of AN are expected to be important for estimating the possible effect of organic molecules on the protein structure and hydration. When a hydrogen bond mediated protein-protein contact is disrupted, sorbate molecules (water or acetonitrile) may differentiate between H-donating and H-accepting fragments of the disrupted contact. Water (H-donor and H-acceptor) is able to solvate both H-accepting and H-donating groups. An H-accepting acetonitrile molecule is expected to prefer H-donating groups, while the remaining H-bond of H-accepting partner will be solvated by water more effectively. Hence, it is expected that acetonitrile molecules are not effective in disrupting the dehydration-induced protein-protein contacts alone. Therefore, no significant acetonitrile-induced structural changes (Figure 9) were observed at the lowest water activity values. At low water content in AN, the residual catalytic activity remains high, equal to ~80% compared with that observed after incubation in pure water [20] (Figure 10).

3.3.2. Intermediate and High Water Activity

According to this model, by penetrating into the initial dried enzyme, water molecules hydrate the polar groups of protein-protein contacts and create new sorption sites at the hydrated (disrupted) contacts. There is a competition between water and acetonitrile molecules for these binding sites. This competition results in considerable differences in hydration and structure compared with that for pure water.

Cooperative decrease in the fraction of the H-bond associated water molecules in the presence of acetonitrile occurs in a narrow water activity range from 0.8 to 0.9 [15]. This cooperative transition results in sharp changes in sorption (Figures 2 and 3) and structural (Figure 9) properties of α-chymotrypsin at water activity above 0.6-0.8. It is worth noting that this transition occurs simultaneously with the decrease in the enzyme storage stability [20] (Figure 10). This means that water hydrogen bond network is sufficient for the expression of the acetonitrile-induced denaturation of α-chymotrypsin.

CONCLUSION

FTIR spectroscopic measurements provides an informative tool in monitoring the inter- and intramolecular processes that occur upon hydration of α-chymotrypsin in the presence of organic solvent.

The obtained results show that the hydration and structure of α-chymotrypsin depends markedly on how enzyme has been hydrated – whether in the presence or in the absence of organic solvent vapor.

An analysis of the concentration dependences shows that, at low water content, the protein-protein contacts in the dried enzyme largely govern its thermodynamic, structural, and functional properties. At high water activity, the chymotrypsin state is determined by the enzyme-organic solvent and enzyme-water interactions resulting in the enzyme denaturation. The interplay of these two factors is responsible for the observed complex shape of the enzyme activity - a_w curves, with an extremum at intermediate water activity values.

REFERENCES

[1] Gregory, R.B. (1995) Protein hydration and glass transition behavior. In: *Protein-Solvent Interactions* (Gregory, R.B., ed.), Marcel Dekker, New York, 191-264.

[2] Rupley, J.A., Careri, G. (1991) Protein hydration and function. *Adv. Protein Chem,* 41, 37-172.

[3] Prestrelsky S.J., Tedeschi N., Arakawa T., Carpenter J.F. (1993) Dehydration-induced conformational transitions in proteins and their inhibition by stabilizers. *Biophys. J,* 65, 661-671.

[4] Van de Weert M., Harvez P.I., Hennink W.E., Crommelin D.J.A. (2001) Fourier transform infrared spectrometric analysis of protein conformation: effect of sampling and stress factors. *Anal. Biochem,* 297, 160-169.

[5] Griebenow K., Klibanov A.M. (1995) Lyophilization-induced reversible changes in the secondary structure of proteins. *Proc. Natl. Acad. Sci. USA,* 92, 10969-10976.

[6] Sirotkin, V.A., Zinatullin, A.N. Solomonov,B.N., Faizullin, D.A., Fedotov, V.D. (2002) Interaction enthalpies of solid bovine pancreatic chymotrypsin with organic solvents: comparison with FTIR-spectroscopic data. *Thermochim. Acta*, 382, 151-160.

[7] Sirotkin V.A., Korolev D.V. (2005) Effect of acetonitrile on the hydration of human serum albumin films: a calorimetric and spectroscopic study. *Thermochim. Acta,* 432, 246-253.

[8] Sirotkin V.A., Solomonov B.N., Faizullin D.A., Fedotov V.D.(2002) Sorption of water and acetonitrile by human serum albumin. *Russ. J. Phys. Chem. A,* 76, 2051-2057.

[9] Sirotkin, V.A. (2005) Effect of dioxane on the structure and hydration-dehydration of α-chymotrypsin as measured by FTIR spectroscopy. *Biochim. Biophys. Acta.*, 1750, 17-29.

[10] Fersht A. Structure and mechanism in protein science: a guide to enzyme catalysis and protein folding. New York: Freeman & Co, 1999.

[11] Lehninger AL, Nelson DL, Cox MM. Principles of Biochemistry. New York: Worth, 1993.

[12] Perrin, D.D., Armarego, W.L.F., Perrin, D.R. (1980) *Purification of Laboratory Chemicals*, Oxford: Pergamon Press.

[13] Kogan, V.B., Fridman, B.N., Kafarov, V.V. (1966) *Equilibrium between Liquid and Vapor. Handbook*, Nauka, Moscow.

[14] Nikol'skii, B. P. (1963). *Chemist's Handbook,* Goskhimizdat: Leningrad, Russia.

[15] Sirotkin, V.A., Solomonov, B.N., Faizullin, D.A., Fedotov, V.D. (2000) IR spectroscopic study of the state of water in dioxane and acetonitrile: relationship with thermodynamic activity of water, *J. Struct. Chem.* 41, 997–1003.

[16] Brunauer, S., Emmet, P.H., Teller, E. (1938) Adsorption of gases in multimolecular layers, *J. Amer. Chem. Soc.* 60, 309-319.

[17] Bull, H.B. (1944) Adsorption of water vapor by proteins. *J. Amer. Chem. Soc.* 66, 1499-1507.

[18] Luscher-Mattli, M., Ruegg, M. (1982) Thermodynamic functions of biopolymer hydration. I. Their determination by vapor pressure studies, discussed in an analysis of the primary hydration process. *Biopolymers* 21, 403-418.

[19] Grdadolnik, J., Marechal, Y. (2001) Bovine serum albumin observed by infrared spectrometry. I. Methodology, structural investigation, and water uptake. *Biopolymers* 62, 40-53.

[20] Sirotkin, V.A., Mukhametzyanov, T.A. (2006) Effect of acetonitrile on the binding of competitive inhibitor proflavin and on the catalytic activity of bovine pancreatic α-chymotrypsin, *Russ. J. Phys. Chem.*, 80, 803–808.

EDITOR'S CONTACT INFORMATION

Dr. Vladimir Sirotkin
A.M. Butlerov Institute of Chemistry
Kazan Federal University
Kremlevskaya str., 18, 420008, Kazan, Russia
Email: vsir@mail.ru

INDEX

A

B

C

D

E

F

G

H

I

K

L

M

S

T

U

V

W

Z